Essential Microbiology for Pharmacy and Pharmaceutical Science

Essential Microbiology for Pharmacy and Pharmaceutical Science

Geoffrey Hanlon and Norman Hodges

School of Pharmacy and Biomolecular Sciences,
University of Brighton, UK

WILEY-BLACKWELL

A John Wiley & Sons, Ltd., Publication

This edition first published 2013 © 2013 by John Wiley & Sons, Ltd

Wiley-Blackwell is an imprint of John Wiley & Sons, formed by the merger of Wiley's global Scientific, Technical and Medical business with Blackwell Publishing.

Registered office: John Wiley & Sons, Ltd, The Atrium, Southern Gate, Chichester, West Sussex, PO19 8SQ, UK

Editorial offices: 9600 Garsington Road, Oxford, OX4 2DQ, UK

The Atrium, Southern Gate, Chichester, West Sussex, PO19 8SQ, UK

111 River Street, Hoboken, NJ 07030-5774, USA

For details of our global editorial offices, for customer services and for information about how to apply for permission to reuse the copyright material in this book please see our website at www.wiley.com/wiley-blackwell.

The right of the author to be identified as the author of this work has been asserted in accordance with the UK Copyright, Designs and Patents Act 1988.

10 06997872

Library of Congress Cataloging-in-Publication Data

Hanlon, Geoff.
Essential Microbiology for Pharmacy and Pharmaceutical Science / Geoff Hanlon and Norman Hodges.
     p. ; cm.
   Includes bibliographical references and index.
   ISBN 978-0-470-66532-9 (cloth) – ISBN 978-0-470-66534-3 (pbk.)
   I. Hodges, Norman A. II. Title.
   [DNLM: 1. Microbiological Phenomena. 2. Anti-Infective Agents–pharmacology. 3. Infection–drug therapy.
   4. Infection–microbiology. 5. Pharmaceutical Preparations. 6. Pharmacological Phenomena. QW 4]
   615.7′92–dc23

                                                                                    2012027340

A catalogue record for this book is available from the British Library.

Wiley also publishes its books in a variety of electronic formats. Some content that appears in print may not be available in electronic books.

Set in 10/12pt Times by Thomson Digital, Noida, India
Printed and bound in Singapore by Markono Print Media Pte Ltd

First Impression 2013

Contents

Preface

As a pharmaceutical scientist or practicing pharmacist, what do we need to know about microbiology?

Whether you are studying for a degree in pharmacy or a related discipline such as pharmaceutical and chemical sciences, you are, first and foremost, pharmaceutical scientists and need to understand the main function of professionals working in this field. The pharmaceutical sciences are a very broad discipline encompassing everything from the basic sciences to the social sciences; however, the bottom line is that pharmaceutical scientists are the only professionals with an intimate knowledge of the production and use of medicines. They are involved at every stage in the production of medicines – from drug discovery, product formulation and manufacture to regulatory control and quality assurance, while pharmacists are specifically involved in overseeing the safe and effective use of medicines in the community and in hospital. However, regardless of where your specific expertise resides, it is important to have a good knowledge of *all* stages in the process.

With this in mind we now need to understand where pharmaceutical microbiology fits into the picture.

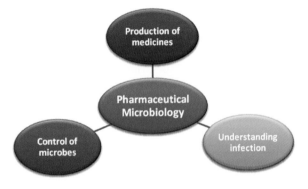

The diagram above illustrates that pharmaceutical microbiology touches on a number of major aspects in the production and use of medicines and we will briefly deal with each of these in turn.

Production of medicines

A large number of medicines and medical devices are manufactured as sterile products – for example, injections and infusions, eye drops and eye ointments, some dressings, catheters and cannulas. Going back in time, there were a number of horrific incidents where products that had not been properly sterilized were given to patients, some of whom subsequently died. It is our responsibility as pharmaceutical scientists to understand the processes of sterilization and aseptic manufacture so that patients' lives are not put at risk.

Even if a medicine is not a sterile product, there are still requirements for it to be manufactured in such a way as to restrict microbial contamination, and quality control procedures including bioburden determination and environmental monitoring are pivotal to achieve this. Many products, such as creams, lotions and liquid oral products, are opened and used on a number of separate occasions. In order to protect the product and the patient we need to incorporate antimicrobial preservatives into these medicines and understand what influences their efficacy.

While we spend a lot of time trying to eliminate microorganisms from our environment and the products we make, it should not be forgotten that they are also responsible for the production of a number of very useful materials. These include antibiotics, steroids, insulin and other recombinant proteins, amino acids, organic acids, enzymes and polysaccharides.

Understanding infection

It is not the role of a pharmaceutical scientist to diagnose infections; that is the job of the clinician, but it is important that we have a clear understanding of viral, fungal and bacterial diseases. The reason this is important is because it critically influences the choice of medication to be used in the management of the disease. Knowing the characteristics of different pathogens, the diseases they cause and the most suitable management protocols is critical for our role in the infection control team. Part of our daily professional routine may be to discuss treatment options with clinical colleagues and we have to be able to speak their language.

We also need to understand how the body reacts to infectious agents so that we can advise patients on how to deal with their symptoms. Many patients will present with minor conditions such as coughs and colds, skin infections, gastro-intestinal problems such as diarrhoea and vomiting and we need to understand what is going on in order to give them the best advice. Moreover, we are seeing increasing numbers of patients with compromised immune systems and so it is necessary to understand why these are more susceptible to infection and how this might be managed.

Control of microbes

An understanding of the control of microbes is relevant in terms of patient treatment and also control of microbes in the environment. From a patient perspective probably the most important role of the pharmacist is to understand the use of antibiotics – what they can do and what they can't. For instance, an antibiotic such as amoxicillin would be of no use in treating a patient with a viral sore throat or a fungal lung infection. Indeed we need to go further and understand why an antibacterial antibiotic such as benzylpenicillin would be ineffective in treating a patient with a Gram-negative bacteraemia. An increasingly important issue and one that will play a major part in the career of any pharmacist or pharmaceutical scientist is that of antibiotic resistance. Many of the antibiotics that we have come to rely upon are losing their usefulness as certain pathogens develop mechanisms to resist their effects. Some clinical isolates of *Pseudomonas aeruginosa*, *Acinetobacter baumannii* and *Mycobacterium tuberculosis* have been found which are resistant to virtually all of the standard antibiotic therapies. How these infections can be managed in the future requires a clear understanding of the basics of microbiology.

Controlling microorganisms in the environment by the use of disinfectants and antiseptics has become increasingly crucial as we understand more about the role of microbes in infection. The use of alcohol gels in hospitals has become the norm and the public has become more aware of the importance of disinfection around the home. Disinfectants are the main tool in deep cleaning procedures applied following outbreaks of infections such as MRSA and *Clostridium difficile* in hospitals. Pharmacists will play a role in implementing these strategies and will need to be in a position to provide advice on all of these scenarios.

The scope of this book

This book covers the microbiology content of the Royal Pharmaceutical Society's syllabus for Pharmacy degrees in the UK which is equally relevant to those studying the pharmaceutical and chemical sciences. It is not intended to give a comprehensive coverage of the whole subject but instead to be an easily digestible outline of the most important features. If the reader requires further information on any of the subjects, the website associated with this book gives examples of where this may be found, and in addition it contains a range of multiple choice questions to allow a reader to check their understanding of the material.

Geoffrey Hanlon and Norman Hodges

Part I
Characteristics of microorganisms

Chapter 1
The microbial world

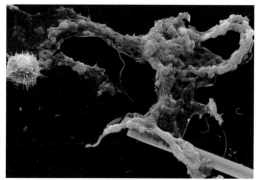

A mixture of bacteria, protozoa and algae in a water sample from a stream

KEY FACTS

- Microorganisms are all around us in enormous numbers and are present both on and within our bodies. Some, termed pathogens, cause disease; others are beneficial and are of commercial importance but the vast majority are harmless.
- Infectious diseases can be caused by agents which are not living microorganisms: prions are simply 'rogue' protein molecules, and viruses usually consist of nucleic acid and protein but have no cellular structure.
- Bacteria represent the simplest living cells. Most of those of pharmaceutical interest can be grown easily in the lab.
- Fungi and protozoa are more complex than bacteria and most of them can exhibit sexual reproduction.
- Relatively few fungi are pathogenic; most are important as contaminants and spoilage organisms in manufactured medicines.
- Protozoa are only of pharmaceutical interest as pathogens; they are not spoilage organisms.

1.1 Microorganisms around us

Microorganisms are present in almost every location and environment on earth. They are in the air, soil and water, on all plants and animals and in such extreme environments as Antarctic ice and rocks 3 km below the earth's surface where the temperature is 60° C or more. Besides growing at extremes of temperature and pH, many bacteria survive and grow in the absence of oxygen; for these bacteria, described as anaerobes, oxygen is toxic. Microorganisms are present, too, in huge numbers and variety. The bacteria in the average human gut are estimated to comprise about 500 different species, and their total number, approximately 10^{14} (one hundred trillion), is about 10 times the number of human cells in the body and more than 10 000 times the human population of the earth. It is impossible to obtain precise data on the relative numbers of harmless and disease-causing (pathogenic) organisms for two main reasons: because new species are being identified all the time, and because of the difficulty of deciding what is harmless and what is not. Organisms that present no threat to a healthy individual might be pathogenic for a person with impaired immunity. Nevertheless, despite the extensive media attention on bioterrorism organisms and the so-called hospital 'superbugs', the harmless bacteria, together with those that are actually beneficial, grossly outnumber the pathogens; one estimate is by a ratio of more than 200 000 to 1.

Essential Microbiology for Pharmacy and Pharmaceutical Science, First Edition. Geoffrey Hanlon and Norman Hodges.
© 2013 John Wiley & Sons, Ltd. Published 2013 by John Wiley & Sons, Ltd.

Table 1.1 Examples of benefits, uses and problems associated with microorganisms.

General benefits and uses	Pharmaceutical applications	Problems and disadvantages
• Essential role in carbon and nitrogen cycles • In brewing, dairy and food industries • In the manufacture of several industrial solvents and other chemicals • As an insecticide • Chemical detoxification • In oil extraction • 'Biological' detergents	• In the manufacture of: antibiotics, steroids, vaccines and many biotechnology products • Used in assays to measure antibiotic concentrations • As biological indicators of sterilization (Chapter 19) • Used in tests to detect metabolic disorders and mutagenicity	• Cause infections • Even harmless species may transmit antibiotic resistance • Even dead bacteria may cause fever (endotoxins) • Contaminate and spoil nonsterile medicines • Cause noninfectious diseases, e.g. gastric ulcers and some cancers

1.2 The benefits of, and problems with, microorganisms

Microorganisms can be essential, passively beneficial or positively useful (Table 1.1). They are essential for the maintenance of life on earth as part of the carbon and nitrogen cycles, for example; without them, dead animals and plants would not decompose and the fertility of soils would fall. Their passive benefits include the protection afforded by probiotic ('friendly') bacteria, which compete with disease-causing species for nutrients and attachment sites on body tissue; they also limit opportunities for harmful bacteria to establish infections in the body by producing antimicrobial chemicals. The practical uses of microorganisms include their long-established rôles in the brewing, dairy and food sectors, and their applications in the pharmaceutical industry, which have multiplied enormously in recent years. Bacteria and fungi have been used since the 1940s to make antibiotics, and since the 1950s in the production of contraceptive- and corticosteroids, but it was the 1980s, a decade which brought major advances in genetic engineering, which saw bacteria used for the manufacture of insulin, human growth hormone, vaccines and many other biotechnology products (see Chapter 20).

Despite their applications in industry and the increasing recognition of their benefits, it is still the case that the main pharmaceutical interest in microorganisms is in killing them or, at least, restricting their contamination and spoilage of medicines. The reasons for this interest are listed in Table 1.1, and although the

first of these – that microorganisms cause infection – is quite obvious, the other problems that microorganisms pose are less well recognized. It is tempting to suppose both that harmless bacteria are irrelevant and that dead bacteria do no harm. Unfortunately, neither supposition is correct: harmless bacteria can carry genes responsible for antibiotic resistance, which they may transfer to disease-causing species, and components of the cell walls of dead bacteria (termed endotoxins) cause fever if they enter the blood stream. Consequently, in order to avoid the risk of fever from residual endotoxins when an injection is administered, it is necessary to ensure that the injection, which must be sterile (free of *living* organisms) anyway, has not been contaminated with high levels of bacteria during its manufacture. However, it is not only sterile medicines where microorganisms can present a problem: the great majority of medicines are not sterile, and the risk here is that the living organisms they do contain may damage the product, either by altering its physical stability or by breaking down the active ingredient.

1.3 The different types of microorganisms

Living organisms are made up of cells of two types: prokaryotic and eukaryotic. Bacteria used to be considered as the only category of prokaryotic cells, but in 1990 a second group, the archaea, were recognized as having equal status to bacteria. Archaea tend to live in inhospitible conditions (high temperatures, extremes of pH or salinity for example) and often possess unusual modes of metabolism, but because no pathogenic archaea have yet been discovered this group will not be considered further. All other organisms are eukaryotic, so the major groups of microorganisms (fungi, protozoa and algae), as well as parasitic worms and mites, and all plants and animals up to and including humans, are eukaryotes. Viruses do not have a cellular structure and so some scientists do not even regard them as living but merely mixtures of complex chemicals; nevertheless, they are indisputably agents of infection and for that reason are usually considered as part of the microbial world.

Recap: the major differences between prokaryotic and eukaryotic cells. (This is not intended to be a full list; several further differences are described in biology textbooks.)

Characteristic	Prokaryote	Eukaryote
Cell nucleus	Do not possess a true nucleus	Have a nucleus surrounded by a nuclear membrane
Nuclear division and reproduction	Mitosis and meiosis are absent so reproduction is asexual	Exhibit both mitosis and meiosis, so reproduction may be sexual or asexual or both depending on species
Genetic variation	Resulting largely from mutations	Resulting both from mutations and the creation of new gene combinations during sexual reproduction
Mitochondria, chloroplasts and ribosomes	Mitochondria and chloroplasts absent; ribosome size is 70s	Mitochondria and chloroplasts may be present; ribosomes larger: 80s
Chemical composition	Do not possess sterols in the cell membrane but do usually have peptidoglycan in the cell walls	Do possess sterols in the cell membrane but no peptidoglycan in the walls

1.3.1 Viruses and prions

Viruses are parasites that infect all kinds of organisms: animals, plants, protozoa and bacteria too. They vary a lot in size and structure, but all contain both nucleic acid and protein; the protein surrounds and protects the nucleic acid core, which may be single-stranded or double-stranded DNA or RNA (Figure 1.1). The largest common viruses are about 300 nm in diameter (e.g. chicken pox virus) and the smallest about 20 nm (e.g. common cold virus), although some which are elongated (e.g. Ebola) may be up to 1400 nm long but very narrow, so none of them can be seen with an ordinary laboratory microscope but only with an electron microscope.

All viruses can only grow inside a host cell and usually the range of hosts is very narrow – often just a single species; rabies is a notable exception. Because they cannot be grown on Petri dishes in the same way as bacteria, they are difficult, time consuming and expensive to cultivate in the laboratory using fertile chickens' eggs or artificially cultured mammalian cells as hosts. Many viruses only survive for a few hours outside their normal host cell, but a few survive much longer and so may, in theory, be present as contaminants of pharmaceutical raw materials of animal origin. However, viruses are relatively susceptible to heat and organic solvents so they are unlikely to arise in materials like gelatin, for example, because of the processing conditions used in their manufacture. Some of the larger viruses have been studied as vectors (carriers) to deliver genes to cells, as in gene therapy for cystic fibrosis for example, but, in general, viruses are, like protozoa, important primarily as pathogens. Although viruses possess genes coding for enzymes to be made by the host cell, the majority contain few, if any, enzymes as part of their

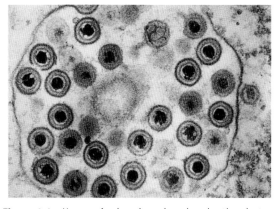

Figure 1.1 Herpes simplex virus viewed under the electron microscope. *Source:* PHIL ID #10231; Photo Credit: Dr. Fred Murphy and Sylvia Whitfield, Centers for Disease Control and Prevention.

structure. One consequence of this is that they are unaffected by the antibiotics used to treat bacterial and fungal infections. This does not mean, though, that it is not possible to create antiviral drugs. One consequence of the HIV/AIDS pandemic is that the number of synthetic antiviral drugs on the UK market increased from fewer than 10 in the mid-1980s to more than 40 by 2010.

Prions represent the simplest infectious agents, which, despite the fact that they are definitely nonliving, are nevertheless usually considered with microorganisms because of their capacity to transmit disease from one person to another. They are similar to viruses in that they have no cellular structure, but differ in that they do not even possess nucleic acids. They are merely atypical mammalian proteins that have the capacity to interact with normal proteins and induce structural changes so that the normal molecules are, in turn, changed into prions that are incapable of fulfilling their normal function. Prions are responsible for fatal, nerve-degenerative diseases termed transmissible spongiform encephalopathies, such as bovine spongiform encephalopathy (BSE; 'mad cow disease') in cattle, and Creutzfeldt–Jakob disease (CJD) in humans. They are particularly stable and difficult to inactivate by disinfectants, gamma-radiation and even by steam-sterilization conditions that far exceed those required to kill the most heat resistant spore-forming bacteria.

1.3.2 Bacteria

Bacteria are responsible for a wider range of diseases than protozoa or fungi, and they were discovered at least 200 years before viruses. For those reasons, and because most of them can easily be grown in the laboratory, bacteria were the most widely studied group of microorganisms throughout much of the nineteenth and twentieth centuries. Typically, they are spherical or rod-shaped cells about 1–10 μm in their longest dimension, so when suitably stained they can easily be seen with an ordinary light microscope (Figure 1.2).

Compared to human cells, bacteria are quite robust: they have a cell wall which protects them against rapid changes in osmotic pressure. Many bacteria will easily survive transfer into water from the relatively high osmotic pressure at an infection site in the body, whereas human cells, without a wall to protect them, would rapidly take in water by osmosis, burst and die. Bacteria are also more tolerant than human cells of wide variations in temperature and pH, and will withstand exposure to higher intensities of ultraviolet light, ionizing radiation and toxic chemicals.

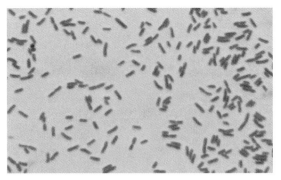

Figure 1.2   The common bacterium *Escherichia coli* (*E. coli*). *Source:* http://commons.wikimedia.org/wiki/File:E_choli_Gram. JPG.

The recap box above highlights the major differences between bacteria (prokaryotes) and eukaryotes, but from a pharmaceutical perspective it is particularly relevant to contrast bacteria with mammals and consider the implications of their differences for the avoidance of microbial contamination and the treatment of infectious diseases. Two of the most fundamental distinctions are that bacteria reproduce asexually whereas mammals exhibit sexual reproduction, and that bacteria may reproduce in as little as 20 minutes but mammalian cells take many hours or days to divide. The cell-division process in bacteria is termed binary fission; this simply involves the chromosome being copied, and the cell enlarging. One copy of the chromosome, together with half the cell contents, becomes separated from the other by the formation of a cross-wall in the cell; a constriction may form which eventually causes the two so-called 'daughter cells' to separate. One bacterial cell doubling every 20 minutes can become over 16 million within 8 hours, and such a large number of cells located together on a Petri dish may become visible to the naked eye as a bacterial colony (Figure 1.3).

The consequence of bacteria reproducing asexually is that they are much more reliant on mutations as a means of producing genetic variation, and the fact that they grow so rapidly means that a mutant can quickly be selected and become the dominant cell type in the population. However, bacteria grow more slowly at an infection site in the human body than on a Petri dish (because they are attacked by the immune system and have to compete for food and oxygen with the body cells) but nevertheless, it is quite possible for antibiotic-resistant mutants to be selected during a course of antibiotic treatment.

The cell structures that are unique to bacteria may be both a benefit and a disadvantage. The cell wall, for

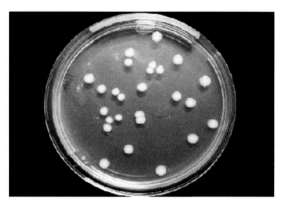

Figure 1.3 Colonies of *E. coli* growing on a Petri dish.

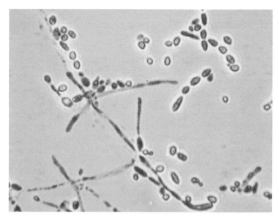

Figure 1.4 Individual cells and filaments of the yeast *Candida albicans* viewed under the light microscope.

example, protects not only against osmotic pressure changes but against drying; consequently, many bacteria survive for long periods in dust. However, the bacterial enzymes that make the cell wall polymers are the targets for a number of important antibiotics, such as penicillins, which achieve their selective toxicity (killing bacteria without harming the patient) simply because human cells do not make cell walls and do not have the enzymes. The same situation applies with respect to ribosomes possessed by bacteria; these are structurally different from those of eukaryotic cells, so antibiotics like tetracyclines and erythromycin interfere with protein synthesis in bacteria but not in humans.

Despite the fact that all bacteria conform to the general description of prokaryotes, they still differ significantly in terms of shape, size and complexity, and these variations have in the past caused problems with classification. Chlamydia and rickettsia are both groups of small, pathogenic bacteria that are obligate, intracellular parasites (meaning that they can only grow within a host cell in a similar way to viruses), whilst mycoplasmas differ from most bacteria in that they do not have a cell wall so they are unaffected by penicillins and other antibiotics that interfere with cell wall synthesis.

1.3.3 Fungi: yeasts and moulds

Fungi are normally thought of as being the toadstools and mushrooms seen on damp, rotting vegetation, but these visible parts of the fungus represent only one stage in their life cycle and other stages involve cells of microscopic dimensions; it is for this reason, and the fact that many fungi never produce structures large enough to be seen with the naked eye, that they are regarded as microorganisms. The word 'fungus' covers both yeasts, many of

which are only slightly larger than bacteria (Figure 1.4), and moulds of the type seen on old food in the fridge.

Yeasts can exhibit sexual reproduction, but more commonly they divide in the same way as bacteria by binary fission or by budding. When growing on a Petri dish their colonies are often similar in appearance to those of bacteria, though usually larger and more frequently coloured (Figure 1.5).

Very few yeasts are capable of causing infection and their pharmaceutical significance is mainly as contaminants of medicines and as spoilage organisms. The term 'mould' is used to describe those fungi that do not produce large fruiting bodies like mushrooms and toadstools. Moulds consist of a tangled mass of multicellular filaments, which, collectively as a colony on a Petri dish, are referred to as a mycelium (Figure 1.6).

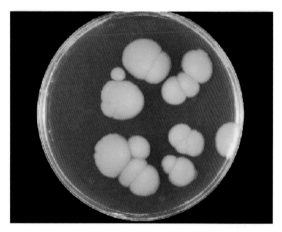

Figure 1.5 Colonies of *Candida albicans* growing on a Petri dish.

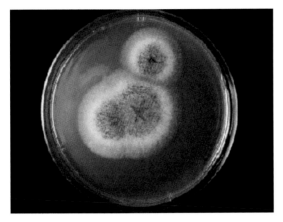

Figure 1.6   Three colonies of *Aspergillus niger* starting to form pigmented asexual spores in the centre of the colonies.

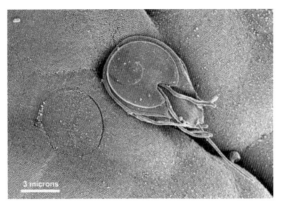

Figure 1.8   A protozoan (*Giardia* species) attached to intestinal epithelium viewed under the electron microscope. *Source:* PHIL ID #11647; Photo Credit: Dr. Stan Erlandsen, Centers for Disease Control and Prevention.

The mycelium is a branched network of tubes called hyphae, which vary in width from 1 to 50 μm and may contain multiple identical nuclei. Like yeasts, moulds are eukaryotes, but this does not mean that sexual reproduction is common. More frequently, moulds reproduce asexually and it is the formation of asexual spores (Figure 1.7) that is often responsible for the characteristic colours seen in many fungal colonies. The periphery of the colony, which is the actively growing region, is often colourless (Figure 1.6). Again, moulds are more significant as contaminants of manufactured medicines than as pathogens, although some are capable of causing severe illness in immunocompromised patients.

Some fungi exhibit different appearances under the microscope depending on their growth conditions. The organism responsible for the infection known as thrush

(*Candida albicans*) often looks like a yeast when grown in the laboratory, but exhibits a pseudomycelium (Figure 1.4) and under the microscope looks more like a mould when isolated from an infection site or from body fluids.

1.3.4 *Protozoa*

Protozoa are single-celled animals that are found in water and soil. The cells are typically 10–50 μm but can be much larger, and are usually motile (Figure 1.8). Some of them feed on bacteria and can be grown in bacterial cultures in the laboratory, but most are difficult to cultivate artificially and they do not, therefore, arise as contaminants of raw materials or manufactured medicines. The great majority are harmless, but a few, such as the organisms responsible for malaria and amoebic dysentery, are capable of causing severe infection, and it is for this reason that they are of pharmaceutical interest.

Table 1.2 summarizes some of the more important distinguishing features of the various groups of infectious agents ranging from the simplest, prions, to the most complex, protozoa.

1.4 Naming of organisms

All microorganisms except viruses are given two names: that of the genus (written with a capital initial letter) followed by the species (with a small initial letter) e.g. *Candida albicans*; it is normally written in italics to indicate that it is a proper name of an individual organism rather than, say, a collection of organisms having similar characteristics e.g. pseudomonads (which

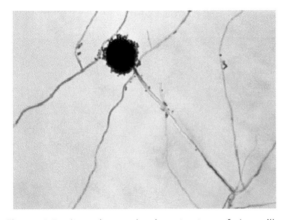

Figure 1.7   Asexual spore-bearing structure of *Aspergillus niger*; the black spherical conidiophore contains many tiny spores that are released when it bursts. *Source:* PHIL ID #3964; Photo Credit: Dr. Lucille K. Georg, Centers for Disease Control and Prevention.

Table 1.2 Distinguishing characteristics of the major groups of infectious agents.

	Cellular structure?	Prokaryote or eukaryote	Genetic material	Laboratory cultivation	Pathogenic potential
Prions	No	Not applicable	No nucleic acids	Only within living organisms	All mammalian prion diseases are untreatable and fatal
Viruses	No	Not applicable	DNA or RNA	Only within living organisms	Most cause active disease[a]
Chlamydia and rickettsia	Yes	Prokaryotes – parasitic bacteria	DNA in a single chromosome but not in a nucleus	Only within living organisms	Many are human pathogens
Bacteria	Yes	Prokaryotes	DNA in a single chromosome but not in a nucleus	Most of those causing human infection can be grown easily	Despite many being pathogens, most are harmless
Fungi	Yes	Eukaryote	DNA in multiple chromosomes in a nucleus	Most can be grown easily	A few are pathogens but the majority are harmless
Protozoa	Yes	Eukaryote	DNA in multiple chromosomes in a nucleus	The majority are difficult to grow in the lab	A few are pathogens but the majority are harmless

[a]But some may exist in a latent form within the host, causing no obvious disease.

describes bacteria similar to *Pseudomonas aeruginosa*), which would not be in italics; coliforms, staphylococci, streptococci and clostridia would similarly be in roman type without an initial upper case letter.

The name of the genus may be abbreviated to a single letter if that abbreviation is unambiguous, so *Escherichia coli* is more frequently written simply as *E. coli*. Names of organisms written in roman type and underlined are still occasionally encountered; this was an old convention by which a typesetter was instructed to set a word in italics, and predates modern word processing, which permits italics to be used directly.

Acknowledgement

Chapter title image: PHIL ID #11715; Photo Credit: Janice Haney Carr, Centers for Disease Control and Prevention.

Chapter 2
Handling and growing microorganisms

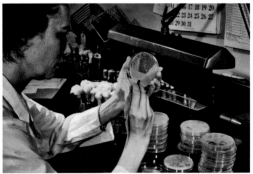

Sampling bacteria from a Petri dish

KEY FACTS

- The word 'sterile' means the complete absence of life, whereas 'aseptic' means a procedure designed to avoid unwanted transfer of organisms from one item or place to another.
- Aseptic procedures can protect the operator from hazardous organisms and the material being handled from contamination.
- Microorganisms are placed in one of four categories according to the infection risk they pose; these are designated hazard groups 1–4, with 4 being the most dangerous. Organisms of pharmaceutical interest are usually in groups 1 and 2.
- Culture media for bacteria usually contain hydrolysed protein as a source of amino acids and B-group vitamins from yeast extract. Media for fungi often have higher sugar concentrations and a lower pH than those for bacteria.
- Microorganisms are normally grown either in liquid media (called broths) or on 'solid' media in Petri dishes. Anaerobic organisms must be grown without oxygen.
- The growth of microorganisms may be affected by a variety of factors including temperature, pH, redox potential, osmotic pressure, nutrient availability and the gaseous environment.

2.1 Sterility and asepsis – what do they mean?

It is obvious that microbial cultures need to be handled in a safe manner in order to avoid the risk of infection, so the important aspects of microbiological safety will be considered in this chapter. But first, it is essential to fully understand the meaning of words used in descriptions of safety procedures. Two of these words, *sterile* and *aseptic,* are commonly used both in relation to safety and in the context of manufacturing and dispensing of medicines. Unfortunately, the words are often misunderstood and used incorrectly as if they mean the same thing – they don't!

Sterile, in a pharmaceutical context, means the complete absence of life. So any medicine or surgical device, or, indeed, any object that is sterile, has no living organisms at all in it or on it. It is an absolute term, so an object is either sterile or it is not; there are no levels of microbial contamination that are so low as to be regarded as insignificant and therefore acceptable. If a medicine is

Essential Microbiology for Pharmacy and Pharmaceutical Science, First Edition. Geoffrey Hanlon and Norman Hodges.
© 2013 John Wiley & Sons, Ltd. Published 2013 by John Wiley & Sons, Ltd.

contaminated with a single organism it is not sterile, so phrases like 'almost sterile' or 'more sterile' should be avoided because they simply display a lack of understanding of the concept.

Several categories of medicines, notably injections and eye products, are required to be sterile, and there are two manufacturing strategies available: the preferred method, known as terminal sterilization, is where the medicine is made from nonsterile raw materials and subjected to a heat, radiation or other sterilization procedure at the end of the manufacturing process. The alternative, used when the product cannot withstand the high temperatures or radiation doses of a terminal sterilization process, is to start with raw materials that are individually sterilized – often by passing solutions of them through bacteria-proof filters – and then mixing them together under conditions that do not allow the entry of microorganisms, followed by filling into presterilized containers; this is termed 'aseptic manufacture'.

Aseptic, therefore, is a word used to describe a procedure that is intended to avoid the unwanted transfer of microorganisms from one object or location to another. It works both ways: it can be a procedure designed to avoid the introduction of organisms into a medicine whilst it is being made, dispensed or administered to a patient, but it also describes procedures for handling hazardous organisms which pose an infection risk to operators. In this case, therefore, the aseptic procedure is intended to keep the organism in its container and avoid it being dispersed into the atmosphere and inhaled, or transferred onto the body of the person handling it. Regardless of whether the intention is to protect the product or protect the operator, aseptic procedures would normally require gowns, gloves, facemasks, disinfectants and the use of safety cabinets or isolators supplied with filtered, decontaminated air.

2.2 Hazard categories of microorganisms

Clearly, microorganisms differ in terms of the infection risk they pose, and many are harmless; others, termed opportunist pathogens, usually only infect individuals with impaired immunity; and some cause incurable, possibly fatal infections. Hazardous organisms need to be contained – that is, handled in laboratories that are designed to protect operators and with facilities which are proportionate to the risk the organism represents. To ensure that laboratory staff are aware of the hazard involved and the degree of laboratory containment

required for any individual organism, the UK government's Advisory Committee on Dangerous Pathogens has assigned microorganisms into one of four categories (designated hazard groups 1–4 with 4 being the most dangerous) in an *Approved List of Biological Agents*, and described the laboratory facilities necessary for the containment of organisms in each group. It is important for persons working in microbiology laboratories in the pharmaceutical industry, hospital pathology laboratories or laboratories attached to hospital manufacturing units to be familiar with the classification scheme and, in particular, with the laboratory design features associated with each category of containment. An organism is classified on the following criteria:

- Is it pathogenic for humans?
- Is it a hazard to laboratory workers (does it survive in the dry state or outside the body for example)?
- Is it readily transmissible from person to person?
- Are effective prophylaxis (vaccines) and treatment (antibiotics) available?

Organisms in group 4 are exclusively viruses, and all of them are readily transmissible pathogens causing untreatable infections with high mortality rates for which there is usually no vaccine. Even some of the notorious bioterrorism pathogens like those responsible for anthrax and bubonic plague are only hazard group 3 because both are bacteria for which there are effective antibiotics. Table 2.1 summarizes the characteristics of each group.

The three classes of safety cabinets all provide operator protection and are designed to minimize the risk of aerosol generation and inhalation of microorganisms; classes II (Figure 2.1) and III additionally provide product protection. The working area in a class III cabinet is totally enclosed and the operator is separated from the materials being handled by a physical barrier, for example integral gloves attached to the front of the cabinet. The design features of the laboratories for each category of containment are too detailed to reproduce here, but are available in a UK Health and Safety Executive book titled *The Management, Design and Operation of Microbiological Containment Laboratories*. There are recommendations on ease of cleaning and absorbency of work surfaces (metal or impervious plastic rather than wood, for example), control of personnel access, operator working space, ventilation, washing facilities, disinfection, laboratory clothing, disposal of waste and accident reporting. Level 3 containment is not commonly encountered in a pharmaceutical setting, and level 4 not at all; containment level 2 would be appropriate for

Table 2.1 Hazard classification of microorganisms by the UK Advisory Committee on Dangerous Pathogens.

Hazard group	Description	Typical organisms	Containment
1	Organisms not normally considered harmful	Lactobacilli (found in milk) and baker's yeast	Open laboratory bench
2	Possible hazard for laboratory staff, but unlikely to spread in the community; antibiotics and vaccines usually available	E. coli, staphylococci and streptococci	Open bench, or class I or II safety cabinets
3	Definite hazard to laboratory staff and may spread in the community; antibiotics and vaccines may be available	The organisms responsible for typhoid, diphtheria, HIV and rabies	Class I or class III safety cabinets[a]
4	Serious hazard to laboratory staff and high risk to the community; antibiotics and vaccines not normally available. Exclusively viral infections	Marburg, Ebola and Lassa fever viruses	Class III safety cabinets

[a] Class II cabinets are designed to protect the product being handled against microbial contamination, and the level of operator protection they afford is not considered adequate for some hazard group 3 pathogens.

organisms reasonably likely to be encountered as contaminants of pharmaceutical materials or organisms used in tests and assays.

Figure 2.1 A microbiologist working with influenza virus using a class II safety cabinet in which the air is passed through filters to remove microorganisms. *Source:* PHIL ID #7988; Photo Credit: James Gathany, Centers for Disease Control and Prevention.

2.3 Sources and preservation of microorganisms

There are several tests and assays described in international standards, pharmacopoeias and regulatory documents for which specified organisms must be used, for example antibiotic assays, preservative efficacy tests and tests for sterility. Usually a particular strain of the organism is required. A strain is a subdivision of a species; it conforms to the standard species description in all major respects but is distinguishable by the possession of a particular property, for example the production of a particular metabolite, enzyme or toxin, or resistance to an antibiotic. Authentic, pure cultures of the test strains are obtainable from national or international culture collections, and strains are identified by the culture collection initials and reference number. The American Type Culture Collection (ATCC) and the UK National Collection of Industrial and Marine Bacteria (NCIMB) are two whose strains are often specified in official methods, but there are others specializing in particular categories of organisms, such as pathogenic bacteria or fungi.

Bacteria and fungi can grow quickly and pass through many generations in a short time, so if a strain is grown in the lab for months or years there is a real possibility of mutants arising with properties that differ significantly from the original cells. If a strain that has changed in this way is used for a test or assay it may not give results that are consistent with those obtained previously, so many test methods direct that the cells to be used must not be more than a specified number of subcultures removed from the original culture collection specimen (subculturing simply means regrowing the organism in fresh, sterile culture medium).

Organisms may be obtained from many other sources and preserved in the laboratory for research or reference purposes. This is common with antibiotic-resistant strains, and the research laboratories of companies developing new antibiotics may keep literally hundreds or thousands of strains with which to compare the effectiveness of a new antibiotic with established ones. It is also common for pharmaceutical companies to keep organisms isolated from their manufacturing areas (these are termed environmental isolates) and from contaminated products in which microorganisms have grown, particularly if they have done so despite the presence of chemical preservatives.

Strains from culture collections are usually supplied in sealed ampoules containing the organism freeze-dried in gelatin, which has to be reconstituted with growth medium when the ampoule is opened (Figure 2.2). Most of the bacteria and fungi of pharmaceutical and medical importance can be stored freeze-dried for many years without significant loss of viability. Alternatively, they can be stored in sealed containers (plastic screw-capped ampoules called cryotubes) either in liquid

nitrogen at $-196°$ C, or in freezers at $-80°$ C. In each case, glycerol (5–15% w/w) is often added to the culture beforehand to protect against freeze-thaw damage.

Many organisms will readily survive storage in a refrigerator at $4°$ C for several weeks or even months, so it is common for bacteria to be kept in this way in Petri dishes sealed with tape, or in screw-capped bottles in which the culture gel was allowed to set at an angle in order to increase the surface area available for growth; these are referred to as slopes or slants (Figure 2.2). If Petri dishes are left unsealed, water evaporates from the culture gel and this may accelerate the death of the organism.

When preserving cultures it is not only important to ensure that the cells are alive but also that they are not contaminated with other species – the purity of the culture is essential. This is confirmed by streaking a liquid culture onto a Petri dish using an inoculating loop (a thin wire attached to an insulated handle; the wire is twisted at the end into a loop of approximately 5 mm diameter so that it will retain a thin film of any liquid in which it is immersed – similar to a child's soap bubble tub!). Streaking is simply a means of spreading a small volume over the surface of the gel in order to dilute it and separate the cells or colony-forming units (often abbreviated to CFU and meaning a group of cells attached to each other so that they give rise to just a single colony after incubation); the operator can then see any contaminant colonies that may differ from the others in terms of shape, size, colour etc. There are several ways in which a Petri dish can be streaked, one of which is shown in Figure 2.3A. The inoculating loop is first heated to redness in a flame to kill any organisms that may already be on it (or a sterile, disposable plastic loop may be used). The Petri dish in Figure 2.3B has been streaked well and indicates that the culture is pure because all of the colonies look alike, but the streaking in Figure 2.3C is unsatisfactory because there is inadequate separation of the individual colonies so that the presence of contaminants may not be recognized.

2.4 Growth media and methods

Some bacteria can grow on simple culture media containing sugar as a carbon and energy source, an ammonium salt as a source of nitrogen for making proteins, and a mixture of minerals like magnesium and iron sulfates. However, such bacteria are in the minority, at least amongst those of interest in pharmacy and medicine, so although simple glucose/salts media are sometimes used for research purposes it is much more common to use so-called 'complex' media (also known as general-purpose

Figure 2.2   A freeze-dried ampoule of *Bacillus subtilis* NCIMB 3610, as received from the culture collection, and an agar slope of the mould *Aspergillus niger*.

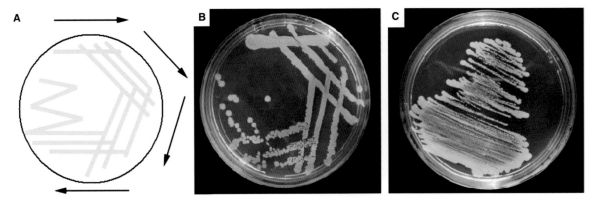

Figure 2.3   Streaking a Petri dish to give isolated colonies and check culture purity (see text).

media). They are complex in the sense that they contain a large number of individual ingredients and the precise composition may vary slightly from batch to batch.

The major ingredients of a complex medium are (in addition to purified water):

- Protein (from meat, milk or soya) which has been hydrolyzed with trypsin or acid to provide the amino acids necessary for growth. Proteins that are only partially hydrolyzed are termed peptones, and peptones produced by the use of trypsin are often called tryptone. Thus tryptone soya medium is one of the most commonly used in routine microbiology and is often specified in pharmacopoeial methods although, confusingly, the pharmacopoeias refer to it as soya bean casein digest medium – a name that is otherwise rarely used.
- A source of B-group vitamins, which enhance microbial growth, so yeast extract is a common ingredient; fat-soluble vitamins are not required.
- Sodium chloride may be included to adjust the osmotic pressure.

A liquid medium containing such ingredients is often termed a 'broth' because of the nutritional similarities to broths used in cooking. In addition to growth in tubes of liquid broth, microorganisms are commonly grown in Petri dishes (often referred to by microbiologists as 'plates') on the surface of a medium solidified by the addition of a gelling agent. These so-called 'solid' media are liquid when hot (so that they may easily be poured into the plates) but the medium sets to a gel upon cooling to room temperature. The gelling agent universally used is agar (a carbohydrate extracted from seaweed) because it cannot easily be digested by bacterial and fungal enzymes and it remains as a gel at 37° C (i.e. body temperature, which is also the preferred growth temperature for most

pathogens) whereas gelatin, a possible alternative, is easily digested and is molten at 37° C. Agar is unusual in that the temperatures at which its gels liquefy and set are not the same: it liquefies at 85° C, but does not set to a gel again until it has cooled to below 40° C. This last property can be an inconvenience, because if an agar medium starts to set whilst Petri dishes are being poured, it is not simply a question of heating by a few degrees to make it more fluid again – it has to be reheated nearly to boiling point. Despite the term 'solid' being used to describe gelled media in Petri dishes, it is important to recognize that they still contain approximately 97% water.

2.4.1 Growing anaerobes

Some microorganisms, termed anaerobes, do not grow on Petri dishes incubated in air; they do not even grow well in an ordinary liquid medium because it contains dissolved oxygen from the air. Anaerobes, therefore, will only grow in reducing conditions – in media with a low redox potential (this term describes whether oxidizing or reducing conditions exist and is measured in mV on a scale having positive and negative values – the larger the negative value the more reducing are the conditions). Air may be removed from liquid media by boiling, but it slowly redissolves after the liquid has cooled so the more common option is to add a reducing agent to the medium; sodium thioglycollate is the usual choice, although sulfur-containing amino acids like cysteine are also used. Anaerobic media often contain redox indicators which change colour according to redox potential rather than pH, for example resazurin, which is pink in oxidizing conditions and colourless when reduced.

In order to check the purity of cultures of anaerobic organisms it is necessary to grow isolated colonies by incubating Petri dishes in an anaerobic jar from which

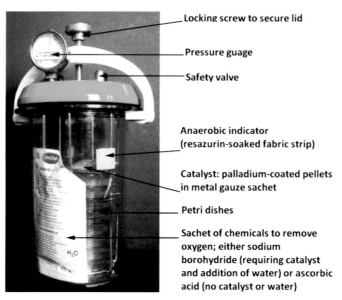

Locking screw to secure lid

Pressure guage

Safety valve

Anaerobic indicator
(resazurin-soaked fabric strip)

Catalyst: palladium-coated pellets
in metal gauze sachet

Petri dishes

Sachet of chemicals to remove
oxygen; either sodium
borohydride (requiring catalyst
and addition of water) or ascorbic
acid (no catalyst or water)

Figure 2.4   An anaerobic jar.

the oxygen has been removed (Figure 2.4). The features of an anaerobic jar are:

- The jar itself, which is usually made of thick polycarbonate with an airtight lid, and is large enough to hold up to 15 standard plates in a stainless steel rack. There is normally a pressure gauge and vent valve in the lid.
- A means of removing oxygen; this is achieved in one of two ways. The traditional method is to combine it with hydrogen by means of a low-temperature catalyst consisting of palladium-coated pellets in a wire gauze sachet that is placed in the jar. The hydrogen is generated *in situ* by addition of water to sachets containing sodium borohydride. The alternative is to combine the oxygen with ascorbic acid (also contained in opened sachets in the jar) – a process that does not require a catalyst.
- A means of confirming that anaerobic conditions have been achieved: normally a fabric strip soaked in resazurin as a redox indicator (see above).

Some pathogens, whilst not being strict anaerobes, nevertheless require gaseous conditions with either reduced oxygen or elevated carbon dioxide levels in order to grow on Petri dishes (e.g. species of *Campylobacter, Streptococcus* and *Haemophilus).* These organisms are usually grown in anaerobic jars and a range of sachets of chemicals are available to achieve the correct gaseous conditions.

2.4.2 *Growing fungi*

Fungi differ from bacteria by tolerating much higher osmolarities and preferring more acid conditions for growth, so fungal media often have high concentrations of sugars and a pH of 5–6. Antibacterial antibiotics are also used in some fungal media, particularly those employed for measuring the levels of fungal contamination in the atmosphere of pharmaceutical manufacturing areas; oxytetracycline and chloramphenicol are used in this way to prevent, or at least minimize, the growth of bacterial colonies.

2.5 The bacterial growth cycle

When bacteria are grown in a container of broth or streaked onto a Petri dish their growth usually follows the pattern in Figure 2.5 (which shows values for both cell concentration and time that would be typical of laboratory-grown cultures). This graph, in which the *logarithm* of the cell concentration is plotted against incubation time, is often seen to comprise four phases: the lag phase, logarithmic (log) phase (also sometimes called the exponential phase), the stationary phase and the decline phase.

For a laboratory culture, the starting concentration would be determined by how heavily the medium was

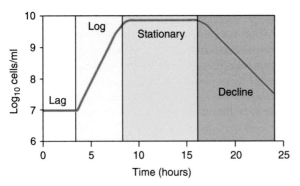

Figure 2.5 Bacterial growth cycle.

inoculated by the operator, but usually sufficient bacteria are added to a container of sterile broth to make the liquid just cloudy, so this would approximate to 10^7 CFU/ml. During the lag phase there is no increase in cell numbers; the bacteria are adjusting to the medium and synthesizing the enzymes and other materials that they will need to grow. Because the cells divide by binary fission (Section 1.3.2) their numbers double in a geometric progression (1, 2, 4, 8, 16... and so forth), which results in a straight line for the logarithmic phase of the plot. During this period the cells are typically breaking down readily digestible carbon sources like glucose, and generating organic acids, for example lactic acid; this is described as primary metabolism (see Chapter 20). Ultimately, the log phase stops due to a shortage of one or more essential nutrients, or to the accumulation of toxic metabolites, such as acids in bacteria or ethanol in yeasts (many of which show the same growth pattern). Typically, aerated cultures of common bacteria like *E. coli*, *Bacillus* species, or pseudomonads will reach a cell concentration approaching 10^{10} CFU/ml at the end of the log phase. The stationary phase is shown in the figure as being of 7–8 hours, although the duration can vary widely depending on the temperature, pH and nutrient status of the culture. Here, the bacteria are undergoing secondary metabolic reactions in which the organic acids are further modified into a wide range of metabolites, several of which, like antibiotics, are commercially valuable (Chapter 20). The rate at which the bacteria die in the decline phase is also very variable, as is the 'final' concentration. In some cases the culture completely dies out, though this may take days or weeks. In the case of bacteria that form spores the final, stable, concentration may be significantly higher than the starting value.

It is possible to calculate the cell generation time during the log phase of growth from the following equation where X is the number of generations achieved during the time required for the population to increase from N_o to N

$$Log_{10}N = Log_{10}N_o + X.Log_{10}2$$

From this

$$X = \frac{(Log_{10}N - Log_{10}N_o)}{Log_{10}2}$$

If, for example, the population increased from 5×10^7 CFU/ml to 1×10^9 CFU/ml in 3 hours

$$X = \frac{(Log_{10}10^9 - Log_{10}5 \times 10^7)}{Log_{10}2} = \frac{9 - 7.699}{0.301}$$
$$= 4.32 \text{ generations in 3 hours}$$

So the mean generation time would be 0.69 hours.

2.6 Environmental factors influencing microbial growth

Several of the factors that influence the rate and extent of microbial growth have already been mentioned in passing in the preceding sections; they include temperature, pH, redox potential, gaseous environment, osmotic pressure and nutrient availability. These will be considered in more detail in later chapters so it is sufficient here just to emphasize that each microorganism will have its preferred optimum for each of the above factors and will grow at a slower rate or to a lower population level when conditions are either side of the optimum. Some of the more important points about environmental effects on microbial growth and survival are listed below.

2.6.1 Water availability and osmotic pressure

All organisms need water to grow, but not necessarily to survive – many, particularly those forming spores, can survive well without water. The amount of 'free' water available for growth in a culture medium (water that is not bound by hydrogen bonding) is indicated by the *Water Activity* (symbol Aw, measured on a scale from 0 to 1, where 1 represents pure water with no dissolved solutes). Bacteria normally require higher water activity values than fungi. The osmotic pressure of a solution will rise as the amount of solute is increased so, generally, solutions containing large

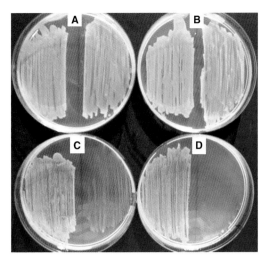

Figure 2.6 A comparison of salt tolerance in *Staphylococcus aureus* (on the left of each Petri dish) and *E. coli* (on the right) grown in media with different concentrations of sodium chloride (%w/v): A, 0.5; B, 2.5; C, 5.0; D, 7.5. Note that *E. coli* fails to grow at all in the highest concentration and only grows weakly at 5%w/v (plate C). The use of high salt concentration is the basis of the selective medium *mannitol salt agar* which is recommended in the pharmacopoeias for the detection of *Staph. aureus*.

amounts of dissolved sugars or salts will have low water activities. However, the two parameters water activity and osmotic pressure are not the same, because osmotic factors are not the only ones that influence water activity. Many organisms grow best at osmotic pressures similar to those in their 'natural' environment, so, for example, many pathogens prefer osmotic pressures similar to those at the sites in the body where they typically cause infections, for example staphylococci grow in relatively high salt concentrations because they are regularly exposed to salt from sweat on the skin (Figure 2.6).

2.6.2 pH

There are microorganisms that can survive in extreme pH conditions, such as hot acid springs in volcanic areas, but survival at extremes of pH is rare amongst organisms of pharmaceutical importance. *Helicobacter* species, the organisms associated with gastric ulcers, will grow at stomach pH (1–3) and *Vibrio cholerae,* the organism responsible for cholera, will grow at pH values between 8 and 9, but most pathogens will grow within a range of about 3 pH units and prefer values near neutrality because the pH of body fluids is usually close to 7.0. Fungi typically prefer, or tolerate, more acid conditions than bacteria.

2.6.3 Temperature

As with water availability, pH and all other environmental conditions, it is important to distinguish what is required for growth from that tolerated for survival. Most pathogens will prefer body temperature (37° C) for growth, and will grow more slowly at slightly lower and slightly higher temperatures too. Usually, though, the upper limit for growth is about 42–44° C because higher temperatures than that result in enzyme inactivation. Some pathogens will grow at refrigeration temperatures, for example *Listeria*, but all organisms require *liquid* water to grow, so the normal freezing point of 0° C is the lower limit. All microorganisms will survive at subzero temperatures provided that damage due to osmotic effects and ice crystals is prevented by cryoprotectant chemicals like glycerol. Survival at higher temperatures varies greatly between species, but spore-forming bacteria tolerate higher temperatures than any other organism – even boiling water.

Organisms may be classified according to their preferred growth temperature ranges: thermophiles are those with an optimum above 45° C; mesophiles have an optimum between 20° C and 45° C; and psychrophiles grow best at a temperature below 20° C.

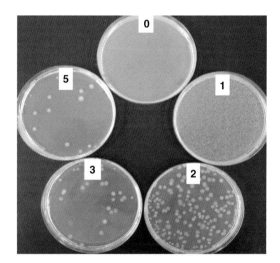

Figure 2.7 The effect of ultraviolet light (UV) on the survival of *Bacillus subtilis* spores. Equal numbers of spores were inoculated and spread over the agar surface of each plate and then exposed to UV for the number of minutes indicated. After this the plates were incubated to allow surviving spores to grow into colonies. On the plates exposed for zero and one minute the number of survivors is so high that the colonies developing after incubation form a complete surface covering of the agar and discrete colonies cannot be seen.

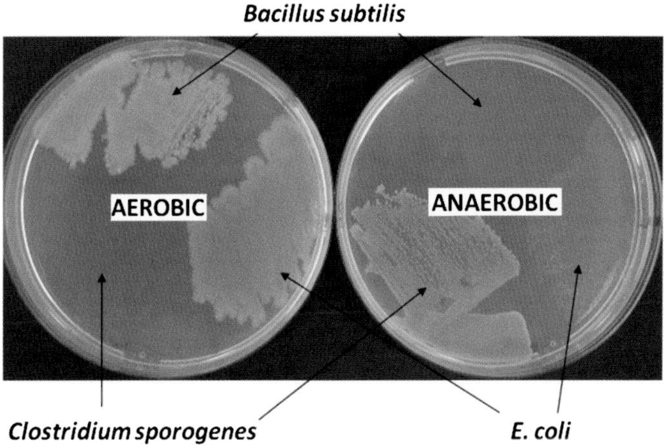

Figure 2.8   A comparison of bacterial growth under aerobic and anaerobic conditions. *Clostridium sporogenes*, a strict anaerobe, will not grow in air, whilst *Bacillus subtilis*, a strict aerobe will not grow under anaerobic conditions. *E. coli*, a facultative anaerobe, grows well in air, but to only a very limited extent without it.

2.6.4 Light, ultraviolet light and ionizing radiation

Microorganisms of pharmaceutical interest do not photosynthesize, so they grow equally well in light and dark. Like all other organisms, though, their DNA is damaged by ultraviolet (UV) light (Figure 2.7), so they may grow much less well if exposed to direct sunlight. Ultraviolet light is used in the pharmaceutical industry to decontaminate air in safety cabinets and for killing microorganisms in purified water to be used for manufacturing medicines (Chapter 19). Microorganisms are killed by X-rays, and killed even more effectively by electron beams and gamma rays, which are used as a means of sterilizing some drugs and, more particularly, medical devices like catheters, cannulas, valves, pacemakers and prostheses (Chapter 19).

2.6.5 Redox potential and the gaseous environment

These aspects have largely been covered in Section 2.4.1 but the following additional points are worthy of note. Most pathogenic microorganisms are either aerobes or facultative anaerobes (Figure 2.8); strictly anaerobic pathogens are less common. An aerobe is an organism that *requires* oxygen to grow, and a facultative anaerobe is one that can grow with or without oxygen but usually grows faster and to higher population levels if oxygen is available. A few organisms are classed as microaerophiles, for example some species of *Helicobacter*, *Campylobacter* and *Streptococcus*: they do need oxygen to grow, but at a concentration lower than that found in the air.

The effects of several antibiotics are influenced by oxygen availability or redox potential. The aminoglycoside antibiotics (such as gentamicin, amikacin and tobramycin) are less effective in reducing conditions, whilst metronidazole is *only* effective when reduced. Because oxygen availability affects the growth rate of so many organisms it is common to find that antibiotics like penicillins and cephalosporins, which only act on growing bacteria, appear more effective in the laboratory than they do in the body. This is because bacteria at infection sites in the body usually grow much more slowly than those in the laboratory, both because they are competing with human cells for the available oxygen and because they are often covered either in slime layers that they manufacture themselves or in mucus secreted by the patient (as in the respiratory tract for example), both of which restrict oxygen diffusion.

Acknowledgement

Chapter title image: PHIL ID #7851; Photo Credit: Dr. Lucille K. Georg, Centers for Disease Control and Prevention.

Chapter 3
Bacterial structure and function

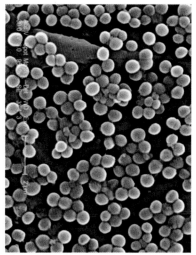

Meticillin-resistant *Staphylococcus aureus* - MRSA

KEY FACTS

- Bacteria are prokaryotic, generally single-celled organisms, which exist in a limited number of morphological forms.
- In the majority of cases their cell walls contains peptidoglycan, a unique bacterial component which confers rigidity to the cell.
- Bacteria can be broadly divided into two main groups – Gram-positive and Gram-negative, based upon the structure of their cell walls.
- One of the components of the Gram-negative bacterial cell wall is lipopolysaccharide which is also known as endotoxin; it is responsible for many of the adverse effects of Gram-negative bacterial infections.
- The bacterial chromosome is a single circular molecule of double-stranded DNA. It is not surrounded by a nuclear membrane but exists in a highly compacted state within the cytoplasm.
- A number of small circular extrachromosomal pieces of DNA may also be found within a cell which will have been inherited from other cells. These may confer on the cell additional attributes such as resistance to antibiotics.
- Some bacteria are able to transform themselves into highly dormant, heat-resistant structures called endospores. This is not a reproductive mechanism but acts as a means of surviving adverse conditions. Bacterial spores are a very important consideration in the design of sterilization procedures.

Chapter 1 has emphasized the fundamental differences between prokaryotic cells and human cells. As we shall see later, this is extremely useful as it allows us to target bacteria with antimicrobial agents which have only a limited effect on us. This chapter looks at the basic structure of bacteria and explains the functions of the various internal components of the cell.

3.1 Bacterial morphology

Although there are some hundreds of thousands of individual bacterial species they exhibit rather limited morphology (structure and shape). Broadly speaking, we can consider those cells whose shape is based upon a rod

Essential Microbiology for Pharmacy and Pharmaceutical Science, First Edition. Geoffrey Hanlon and Norman Hodges.

and those based upon a sphere. Below are some examples with typical dimensions.

3.1.1 Rod-shaped cells

Bacillus

Rod-shaped cells whose dimensions vary depending upon the species. Note the potential confusion in terminology here. The term *Bacillus* is a genus name while the word bacillus also describes a shape. *Escherichia coli* cells are typically 1.0 μm in diameter and 2–3 μm in length, while *Bacillus* species are approximately 2 μm in diameter and up to 7 μm in length.

Vibrio

Rigid curved cells, a typical example of which is *Vibrio cholerae* (the causative agent of cholera). Dimensions are 0.5 μm in diameter by 2 μm in length.

Spirochetes

Thin, flexibly coiled cells e.g. *Treponema pallidum* (the causative agent of syphilis) which has dimensions of 0.5 μm diameter by 5–500 μm in length.

Spirillum

Cells made up of rigid spirals with variable numbers of turns. Examples include *Campylobacter jejuni* (major cause of food poisoning) and *Helicobacter pylori* (implicated in the formation of gastric ulcers). Dimensions are typically 0.5 μm in diameter with variable length up to 60 μm.

Filamentous

Some cells grow as slender, nonseptate, branching filaments, which resemble filamentous fungi rather than typical bacteria. Examples include the streptomycetes, which are soil bacteria responsible for the production of a number of important antibiotics, for example *Streptomyces venezuelae* 0.5–2 μm diameter with variable length.

Fusiform bacilli

'Cigar-shaped' bacteria. *Fusobacterium* species are normal inhabitants of the mouth, gut and female genital tract. They cause a range of infections including, sinusitis, otitis media and dental infections.

3.1.2 Spherical cells (coccus shaped)

These vary not so much in shape as in degree of aggregation and diameter. The extent and the manner in which they aggregate are determined by the plane of cell division and the strength of adhesion between cells after division.

Staphylococcus

These are found as irregular clusters of spherical cells which are said to resemble a bunch of grapes. Cell division takes place in a number of planes with a high degree of adhesion. *Staphylococcus aureus* is often found as part of the normal microflora of the skin and nostrils but can also give rise to wound and other infections, which can be life threatening. Cells are typically 1 μm in diameter.

Streptococcus

These are spherical or slightly oval cells which occur usually in chains. They divide in one plane only and the degree of adhesion between cells is not very strong, hence chains of cells are easily disrupted. An example is *Streptococcus pyogenes*, which can cause sore throats and also skin and soft tissue infections. The cells are approximately 1–2 μm in diameter.

Diplococcus

These are small cocci or oval cells which occur typically in pairs, usually joined along their longest axis and with adjacent sides flattened. An example is *Neisseria gonorrhoea* (often called the gonococcus- which is the causative agent of gonorrhoea).

Tetrad

This is the name given to bacteria which are typically found in clusters of four cells, for example *Gaffkya* species.

Sarcina

The name given to bacteria found in clusters of eight cells, for example *Sarcina* species.

Pleomorphic

These cells exhibit a variable morphology depending upon how they are grown. For example *Lactobacillus* species can grow either as a slender rod-shaped cell or a coccobacillus depending upon culture conditions.

3.2 The cell wall

The structure of a typical bacterial cell is shown in Figure 3.1. One of the most important structures of the cell and one of the things which sets it apart from mammalian cells is its cell wall. If mammalian cells such as erythrocytes are placed in a hypotonic solution they will swell due to the uptake of water by osmosis and then burst (a process known as haemolysis). If bacterial cells are placed in a hypotonic solution they will not burst because of the presence of their rigid cell wall. This cell wall is also responsible for giving them their characteristic shape illustrated above.

However, not all bacteria have the same type of cell wall, and this was first discovered by Christian Gram in 1884. Bacterial cells subjected to a differential staining procedure known as the Gram stain (see text box) were seen under the microscope either as a blue violet colour

or a pink colour. The former were named Gram-positive cells and the latter Gram-negative cells.

> ## Gram staining procedure
>
> 1. Fix bacterial smear onto a glass slide using heat.
> 2. Stain with crystal violet solution – the cells appear blue/violet colour.
> 3. Fix with Gram's iodine (I/KI solution) – the cells same colour as above.
> 4. Decolourize with alcohol or acetone – some cells appear blue/violet, others colourless.
> 5. Stain with safranin solution (red colour).
> 6. Wash with water – Gram-positive cells appear blue/violet colour under microscope, Gram-negative cells appear pink/red colour.

The structures of the two different types of cell wall are shown in Figure 3.2. It can be seen that the Gram-positive cells are much simpler and the predominant component of the cell wall is a polymer called peptidoglycan. The Gram-negative cells are more complex and have a thinner layer of peptidoglycan surrounded by a lipid bilayer comprising lipopolysaccharide and phospholipid. This additional outer layer makes the Gram-negative cells less easily penetrated by drug molecules (including several important antibiotics) and accounts for many of the properties found in these cells.

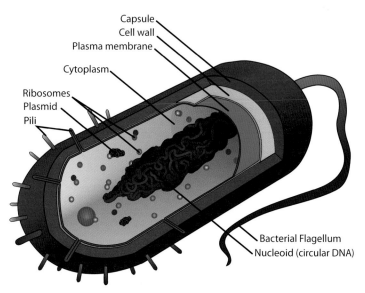

Figure 3.1 Structure of a typical bacterial cell. *Source*: http://commons.wikimedia.org/w/index.php?title=File:Average_prokaryote_cell_en.svg&page=1.

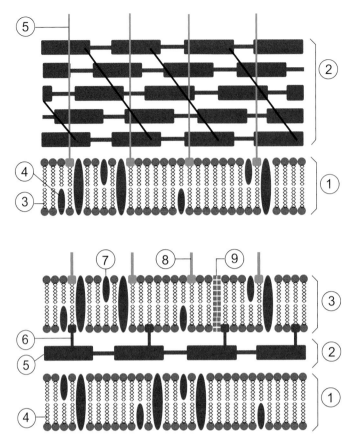

Figure 3.2   Bacterial cell walls. Top: Gram-positive cell wall. 1-cytoplasmic membrane, 2-peptidoglycan, 3-phospholipid, 4-protein, 5-lipoteichoic acid. Bottom: Gram-negative cell wall. 1-inner membrane, 2-periplasmic space, 3-outer membrane, 4-phospolipid, 5-peptidoglycan, 6-lipoprotein, 7-protein, 8-LPS, 9-porins. *Source*: http://commons.wikimedia.org/w/index.php?title=File:Bacteria_cell_wall2.svg&page=1.

3.2.1 Features of peptidoglycan

- They are made up of linear polysaccharide chains (glycan strands) up to 200 disaccharide units in length, cross-linked by short peptide chains.
- Glycan strands are composed of alternating units of N-acetylglucosamine and N-acetylmuramic acid joined by β1–4 glycosidic bonds.
- Peptidoglycan comprises one large continuous molecule surrounding the cell rather like a net.
- The carboxyl group of each N-acetyl muramic acid is attached via a peptide bond to a chain of four amino acids.
- The amino acids which comprise this chain vary but in Gram-negative cells a typical sequence is: L-alanine; D-glutamic acid; *meso* diaminopimelic acid; D-alanine
- In Gram-positive cells the *meso* diaminopimelic acid is replaced by L-lysine.
- The tetrapeptide chains on adjacent glycan strands are joined by peptide bonds to give a cross-linked polymer.

The structure of peptidoglycan is shown in Figure 3.3. This is a generic structure and there are variations between different cells particularly with respect to the cross-linking groups.

3.2.2 Antimicrobial agents acting on peptidoglycan (see Figure 3.4)

Enzymes (for example, lysozyme, lysostaphin):

- Many body fluids such as tears and saliva contain the enzyme lysozyme as a protection against invading bacteria.
- Lysozyme breaks linkages between N-acetyl muramic acid and N-acetylglucosamine.
- Their activity is carefully controlled normally to allow for selective hydrolysis when required.
- On death of the bacterial cell these lytic enzymes are activated and the peptidoglycan of the cell wall is degraded.

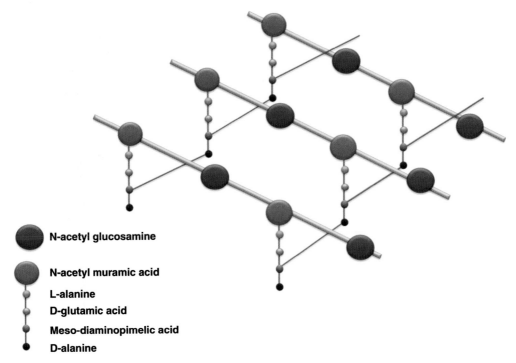

● **N-acetyl glucosamine**

● **N-acetyl muramic acid**

○ **L-alanine**

● **D-glutamic acid**

● **Meso-diaminopimelic acid**

● **D-alanine**

Figure 3.3   General layout of the structure of peptidoglycan.

Antibiotics (for example, penicillins, cephalosporins, bacitracin, vancomycin, teicoplanin):

- β-lactam antibiotics (penicillins and cephalosporins) act by preventing the cross-linking of peptidoglycan during synthesis. There is no effect on intact peptidoglycan, therefore they only act on growing cells.
- Bind to penicillin-binding proteins (PBPs), which are carboxypeptidases and transpeptidases responsible for final stages of cross linking.

- Enzyme inhibition leads to a build up of precursors and release of autolytic enzymes. They form spheroplasts (spherical cells lacking all or part of the cell wall: Figure 3.5) and lysis of the cells results due to fragile outer cell walls.
- MRSA (meticillin resistant *Staphylococcus aureus*) has an additional PBP with lower binding affinity for β-lactams.
- Bacitracin interferes with a lipid carrier responsible for transporting cell-wall precursors across the membrane from the cytoplasm.

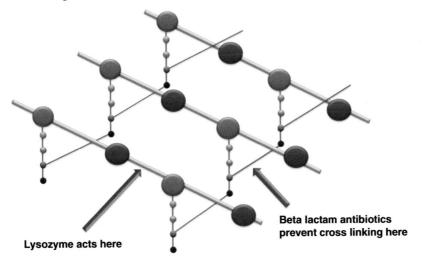

Lysozyme acts here

Beta lactam antibiotics prevent cross linking here

Figure 3.4   Action of lysozyme and beta lactam antibiotics on peptidoglycan.

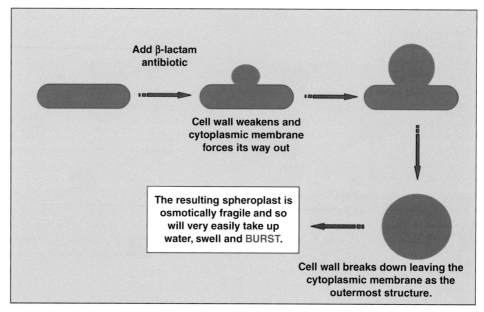

Figure 3.5   The effect of beta lactam antibiotics on a bacterial cell.

- Glycopeptides are large polar molecules unable to penetrate the Gram-negative outer membrane. They are only active against Gram-positive bacteria and act by binding to the end of peptide chains, preventing cell-wall growth.

3.3 Teichoic acids

Teichoic acids are found only in Gram-positive cell walls and constitute up to 45% of the wall of *Staphylococcus aureus*. They do not confer any extra rigidity to the cell wall, but being acidic in nature they may function as sequestering agents extracting essential cations from solution. They are believed to regulate the activities of amidases and glycosidases, which are involved in cell-wall synthesis and also act as adhesins regulating attachment to surfaces. They have also been shown to play a role in inflammation (causing host cells to release inflammatory agents such as cytokines).

3.4 Lipopolysaccharide (endotoxin)

Gram-negative bacteria retain the ability to kill and injure humans even after the cells have died. Suspensions of cells autoclaved for 1 hour at 121°C will be killed and the solution will be sterile. The resulting solution, if injected into a patient, could cause high fever, circulatory collapse and death. Patients with Gram-negative septicaemia (bloodstream infection) are liable to suffer from toxic shock even if given antibiotics. The toxic component is the lipopolysaccharide found in the Gram-negative cell envelope, commonly called endotoxins or pyrogens. It is therefore important to ensure that injections are not just sterile but also free from endotoxins.

A lipopolysaccharide molecule is made up of three parts:

- An outer region, which is a type-specific O-antigen extending into the external environment.
- A core region which anchors the O-antigen to the membrane.
- The lipid A region, which is similar to phospholipid. This sits inside the membrane and all the significant pharmacological activity of LPS is due to this region (see below).

3.4.1 Toxic effects of endotoxins

Pyrogenicity
- Body temperature is regulated by the thermoregulatory centre in the hypothalamus of the brain and endotoxin interferes with this control.
- It does not act directly but causes release of endogenous pyrogen from macrophages in the body.

Cardiovascular effects
- Initially endotoxin causes hypertension but this is followed by progressive and profound hypotension. Shock and death may follow.

Generalized toxic effects:
- Endotoxin stimulates the release of cytokines (such as interleukins and kinins) from macrophages and other cells.

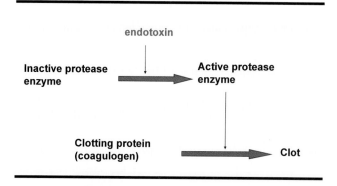

Figure 3.6   Mechanism of LAL clotting test for endotoxin.

- Most of the actions of endotoxin can be attributed to the release of these endogenous mediators.

3.4.2 Limulus amoebocyte lysate (LAL) test for the detection of endotoxins

The horseshoe crab (*Limulus polyphemus*) is a living fossil which has remained unchanged since the time of the dinosaurs. It is a large creature 30 cm across with an armour-plated shell and it devours large quantities of shell-fish – hence it is not liked by fishermen. The creature is seldom seen and lives in deep waters off the coast of North America, Canada and southeast Asia. Every spring they come ashore in their hundreds of thousands to lay eggs.

It has been known for many years that the blood of *L. polyphemus* forms a solid clot when withdrawn from the animal. We now know that coagulation is caused by the presence of contaminating bacterial endotoxin reacting with the contents of circulating blood cells (amoebocytes). The crabs are harvested and their blood withdrawn under sterile conditions. The crabs are not harmed by this and are returned to the sea. The amoebocytes are collected and lysed and the contents formulated into LAL reagent. The clotting reaction is very sensitive and can detect nanogram quantities of endotoxin (see Figure 3.6).

3.5 Cytoplasmic membrane (see figure 3.7)

Cytoplasmic (plasma) membranes of Gram-positive and Gram-negative cells are similar. They are made up of protein (60–70%); lipids/phospholipids (20–30%)

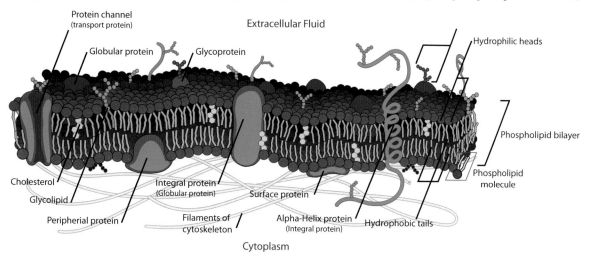

Figure 3.7   Diagram of a bacterial cytoplasmic membrane. *Source*: http://commons.wikimedia.org/wiki/File:Cell_membrane_detailed_diagram_en.svg.

and a small amount of carbohydrate. Sterols such as cholesterol and ergosterol are absent in bacteria but are present in eukaryotic cells where they provide rigidity.

Prokaryotic membranes are very fluid – stabilized by hydrogen bonding and magnesium and calcium ions. The membrane acts as an osmotic barrier allowing only small molecules to enter the cytoplasm. Larger molecules can only enter if their entry is mediated by specific transport proteins and, being hydrophobic, the membrane is impermeable to hydrogen ions.

Oxidative phosphorylation occurs in the membrane. Redox couples (a reducing species and its corresponding oxidized form) pass electrons down the series to oxygen as the final electron acceptor. Energy is used to pump protons out of the cell and this gives a proton and charge gradient across the membrane – known as the proton motive force. Protons are allowed back via a transporter system linked to ATP production.

Bacterial cells establish an internal osmolarity in excess of the environment in order to maintain a constant internal pressure. The cell responds to changes in external osmotic stress by altering the internal concentration of solutes. Solutes include potassium ions, glutamate, glycine betaine and various sugars. Potassium is an important intracellular cation and cells can accumulate high concentrations in excess of the environment (for example staphylococci up to 30x higher).

A large number of general antimicrobial agents (including surfactants) act on the cytoplasmic membrane. They cause loss of membrane integrity, leakage of intracellular contents, disruption of enzyme function and uncoupling of proton motive force.

3.6 Inclusion bodies (storage granules)

Bacteria often accumulate materials within storage granules inside the cell. These can take a variety of forms as shown below:

- Volutin granules
  - are used for storage of phosphorus and energy;
  - consist of polymetaphosphate;
  - accumulate at the end of active growth.
- Glycogen granules
  - are food and energy stores;
  - accumulate under conditions of nitrogen starvation;
  - can account for up to 50% of the dry weight of a cell.

- Lipid granules
  - consist of poly-β-hydroxybutyric acid;
  - are storage products;
  - occupy a large volume in old cells.
- Sulfur/iron granules
  - some cells can accumulate magnetite (Fe_3O_4);
  - impart a permanent magnetic dipole to a cell.

3.7 The bacterial chromosome

The chromosome contains the genetic information necessary for functioning of the cell. It is a single circular molecule of double-stranded DNA approximately 1000 times as long as the cell itself. It exists in a highly folded state and this supercoiling is brought about by topoisomerase enzymes. In prokaryotic cells the chromosome is not surrounded by a nuclear membrane but exists free in the cytoplasm condensed into areas called chromatin bodies. In a feat of extraordinary dexterity the chromosome may duplicate itself every 20 minutes during exponential growth.

The role of the DNA is to act as a source of information for the synthesis of proteins encoded into a sequence of nucleotides. The cell can access this information by firstly transcribing the coded information into RNA by the action of the enzyme RNA polymerase and this is then translated into a peptide sequence by the action of the ribosomes (see box below). The nucleotide sequences on the DNA are collected into discrete areas called genes each with their own control elements.

Prokaryotic genomes are very small with very little space between the genes. *E. coli* has approximately 4000 genes and the length of the DNA molecule is about 1mm. On the other hand a yeast cell has approximately 6000 genes in a genome three times the size of *E. coli.*

Transcription and translation

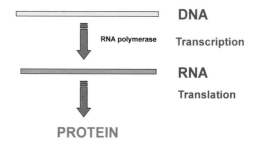

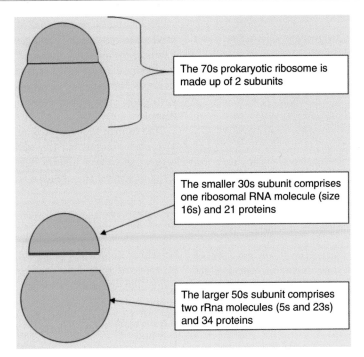

The 70s prokaryotic ribosome is made up of 2 subunits

The smaller 30s subunit comprises one ribosomal RNA molecule (size 16s) and 21 proteins

The larger 50s subunit comprises two rRna molecules (5s and 23s) and 34 proteins

Figure 3.8 Structure of prokaryotic ribosomes.

3.8 Plasmids

Cells can contain additional genetic elements such as plasmids. These are autonomously replicating, extrachromosomal, circular pieces of double-stranded DNA. They vary in size from 1000 to 200 000 base pairs (the *E. coli* chromosome has 4 million base pairs) and encode for accessory functions conferring advantages to the cell, for example the production of toxins, pili, bacteriocins, siderophores and enzymes responsible for antibiotic resistance.

Plasmids replicate faster than the main bacterial genome and so cells usually contain multiple copies. All plasmids contain the information required for replication but some also contain information for cell to cell transfer (called conjugative plasmids). Plasmids may contain multiple genes encoding antibiotic resistance elements and be able to transfer between cells. What is worrying is the ability to transfer plasmids between species, because a relatively harmless bacterium found in our gut, for example. can then donate antibiotic resistance genes to a pathogenic species.

3.9 Ribosomes

Protein synthesis is carried out by the ribosomes which are complex structures approximately 18 nm in diameter.

They have a molecular weight of about 2500 kDal and those from bacteria have a sedimentation coefficient (which is a function of size) of 70s. Those found in eukaryotic cells are slightly larger and have a sedimentation coefficient of 80s. The bacterial ribosomes can dissociate into a large and a small subunit with sedimentation coefficients of 50s and 30s respectively (see Figure 3.8).

Bacterial protein synthesis starts with the interaction between a 30s ribosomal subunit, mRNA and a tRNA (attached to the amino acid formylmethionine). The 50s ribosomal subunit then attaches and the whole ribosome moves along the mRNA chain reading the sequence of nucleotides and constructing the peptide chain dictated by the mRNA.

3.10 Fimbriae (pili)

Fimbriae are common on Gram-negative bacteria but uncommon on Gram-positive. Most of the fimbria is made up of major structural protein. The minor tip protein gives rise to variability and leads to antigenic variation and adhesion variability. *Neisseria gonorrhoea* possesses fimbriae, which facilitate binding to the urogenital epithelium. The cell can switch production from one type of fimbria to another to evade host responses.

Type I fimbriae (common fimbriae) are chromosomally mediated and found in *E. coli* and other enteric

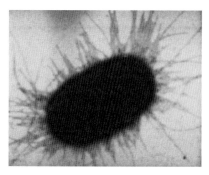

Figure 3.9   Electron micrograph of bacterium with multiple fimbriae. *Source:* Gross L (2006) Bacterial fimbriae designed to stay with the flow. PLoS Biol 4(9): e314. doi:10.1371/journal. pbio.0040314, © 2006 Public Library of Science.

bacteria including *Salmonella* species. These assist the cell in colonizing the large intestine. In enterotoxigenic *E. coli* (see Figure 3.9) the production of fimbriae is plasmid-mediated and these structures allow colonization of the small intestine. The K88 pilus antigen of some *E. coli* causes diarrhoea in pigs but not in other animals as a result of specific cellular adhesion.

3.11 Capsules

Many Gram-negative bacteria and some Gram-positive bacteria produce extracellular polysaccharides (EPS), which may take the form of a discrete capsule or a more generalized layer of slime (Figure 3.10). The EPS can function in:

- adherence (for example, *Streptococcus mutans*);
- resistance (to biocides, desiccation and macrophages);
- virulence (*S. pneumoniae*).

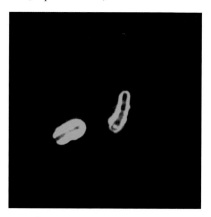

Figure 3.10   Photomicrograph showing *Bacillus anthracis* cells surrounded by a capsule. Visualized using fluorescent dye tagged to capsular antibody. *Source:* Oregon State Public Health Laboratory ID #1888; Photo Credit: Larry Stauffer, Centers for Disease Control and Prevention.

Streptococcus pneumoniae depends upon its polysaccharide capsule for pathogenicity. In the absence of a capsule the infecting dose is increased 10,000 times. The biochemical nature of the capsule can affect virulence, for example two types of encapsulated pneumococcus are pathogens: Type 3 gives severe disease; Type 30 gives mild disease.

The human host responds very well immunogenically to protein antigens but less well to polysaccharides and, hence, the capsule shields cells from detection by the immune system. The capsule is referred to as the K antigen.

3.12 Flagella

Flagella are long whiplike appendages that enable bacteria to move (Figure 3.11) and cells lacking flagella are nonmotile. They are made up of repeating units of a simple protein called flagellin and the numbers of flagella per cell are variable. The flagella are antigenic and are given the name H antigen. The structure does not beat or wave but it is a rigid helix, which rotates like a propeller up to 300x per second. Using this mechanism the bacteria can travel 200 times their own length per second.

Motile bacteria have two modes of movement:

- **Swimming**
  - Directed by the flagella and cause the bacterium to move in a straight line.

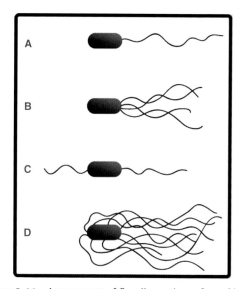

Figure 3.11   Arrangement of flagella on the surface of bacterial cells. (A = monotrichous; B = lophotrichous; C = amphitrichous; D = peritrichous). Image by Mike Jones; http://commons.wikimedia.org/wiki/File:Flagella.png.

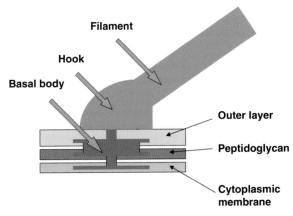

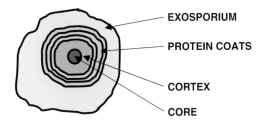

Figure 3.13   Cross sectional diagram of bacterial endospore.

Figure 3.12   Diagram of bacterial flagellum showing basal body, hook and filament.

characteristic morphological and biochemical changes. It is important to understand that a single cell gives rise to a single spore due to internal reorganization. The end result is a multilayered dormant spore as shown in Figure 3.13.

- **Tumbling**
  - Bacterium rotates around its own axis in a random manner but does move from that spot.

Figure 3.11 shows the various arrangements of flagella found on the surface of bacterial cells. The flagellum originates in the cytoplasmic membrane and is firmly anchored in the cell wall (see Figure 3.12). ATP synthesizing systems within the membrane provide energy to drive the action of the flagellum.

3.13 Bacterial endospores

Most vegetative (actively growing) bacteria exhibit the normal growth cycle as shown in Chapter 2. With the advent of nutrient depletion or accumulation of toxic waste products the cell population begins to die. However, a limited number of bacteria can differentiate into highly resistant endospores and so survive extreme environments. The two main endospore-forming genera are:

- *Bacillus* species (aerobic Gram-positive rods);
- *Clostridium* species (anaerobic Gram-positive rods).

The trigger for sporulation to begin is usually nutrient starvation. Sporulation is *not* a reproductive process but a survival mechanism. Spores exhibit extreme resistance to heat, disinfectants, radiation and desiccation and they are highly dormant structures – which can exist in a state of cryptobiosis (where metabolic activity is reduced to an undetectable level). The spores can however, germinate very rapidly when favourable conditions are restored.

The sporulation process takes about seven to eight hours under ideal conditions and each stage is associated with

3.13.1 Why spores are important

- Extreme resistance makes them difficult to eradicate from pharmaceuticals and food.
  - Sterilization processes are designed for the elimination of spores.
- A number of species are dangerous pathogens.
  - *Bacillus anthracis*   **anthrax**
  - *Clostridium tetani*   **tetanus**
  - *Clostridium perfringens*   **gangrene**
  - *Clostridium botulinum*   **botulism**
- Issue of dormancy (longevity in the environment) makes them a problem with respect to germ warfare.
- Spore-forming bacteria are commercially important as producers of antibiotics – for example bacitracin, gramicidin and polymyxin – and of insecticides.
- They are a simple example of cellular differentiation.

3.13.2 Longevity of spores

- Spores have been isolated from lake sediments 500–1000 years old and revived within hours.
- Anthrax spores found sealed in jars known to be over 1300 years old were still viable.
- Spores found in the gut of an extinct bee trapped within amber germinated rapidly in the laboratory. The bee was thought to be 25 to 40 million years old.

3.13.3 Heat resistance of bacterial endospores

The extreme heat resistance of bacterial endospores has long been a fascinating phenomenon for microbiologists. Early theories postulated that the whole

Table 3.1   Heat resistance of endospores from different species.

Organism	Time taken to kill at 100°C
Most vegetative bacteria	Seconds
Bacillus subtilis	15 minutes to several hours
Bacillus anthracis	10 to 15 minutes
Geobacillus stearothermophilus	6 hours
Clostridium tetani	60 minutes
Clostridium perfringens	5 to 10 minutes
Clostridium botulinum	6 to 7 hours

spore must be highly dehydrated although this was quickly shown not to be true. A unique spore component, dipicolinic acid, (DPA) complexed with calcium ions was originally also thought to be involved but this was also shown to be incorrect. It is now thought that only the core is highly dehydrated and this is brought about by the cortex squeezing water out.

Spores of different species vary in their ability to withstand heat and Table 3.1 gives examples of the heat resistance of bacterial endospores. The extreme heat resistance of *Geobacillus stearothermophilus* has led to its use as a biological indicator in steam sterilization processes (Chapter 19).

3.13.4 Activation, germination and outgrowth

An endospore will remain in a dormant state until it encounters an environment favourable for growth. The transformation of a dormant, highly resistant endospore into a fully metabolizing vegetative cell takes place via a series of different steps – activation, germination and outgrowth:

- **Activation**. (This can be likened to the alarm clock waking you in the morning. There is no guarantee that it will result in you actually getting out of bed!)
  - It is the process of breaking dormancy of the spore.
  - It is reversible.
  - Processes include heat shocking and exposure to specific chemicals.
- **Germination**. (The alarm clock has woken you and you can't get back to sleep, so you might as well get out of bed.)
  - The process is irreversible and is the change from a dormant spore to a metabolically active cell.
  - Germination is associated with a number of events:
    - loss of heat resistance;
    - release of DPA;
    - decrease in optical density;
    - loss of refractility.
- **Outgrowth** (now you're up and about and fully active).
  - Development of a vegetative cell from a germinated spore.

Acknowledgement

Chapter title image: PHIL ID #10046; Photo Credit: Janice Haney Carr, Centers for Disease Control and Prevention.

Chapter 4
Mycology: the study of fungi

A *Trichophyton* species: they infect skin, hair and nails

KEY FACTS

- Fungi are eukaryotic cells and are mainly saprophytes (feeding on decaying organic material) and growing at temperatures below 25° C.
- They are responsible for spoilage of foods and medicines and so their growth needs to be controlled.
- Very few are pathogenic to humans, but those infections which do arise are usually persistent and difficult to treat.
- Some fungi are capable of producing materials which are extremely useful to us in the food, chemical and pharmaceutical industries.

Mycology is the name given to the study of fungi. This is a huge subject area but the vast majority of it is of no importance to those of us studying the pharmaceutical sciences. There are some areas which have an impact on our field and so these are the parts where we need to have a working knowledge of the subject. The parts we need to know something about are those where the fungi cause spoilage of products we make, where they cause disease and where the cells have some role in the production of pharmaceutically useful materials.

This chapter will comprise a general overview of the subject and introduce the main fungi of interest. More details on product spoilage and preservation, industrial aspects of mycology and the treatment of pathogenic fungi will be dealt with in other chapters.

4.1 Definitions

- **Fungus** is a general term and is used to describe both unicellular yeasts and multicellular moulds.
- **Moulds** are multicellular fungi usually having a branching, filamentous structure.

4.2 Main characteristics

Yeasts and moulds are eukaryotic organisms and thus they possess a nucleus, endoplasmic reticulum, Golgi apparatus, mitochondria, nucleolus and so forth. The fungal cell wall is composed of various polysaccharides, including chitin, but is markedly different from those

Essential Microbiology for Pharmacy and Pharmaceutical Science, First Edition. Geoffrey Hanlon and Norman Hodges.
© 2013 John Wiley & Sons, Ltd. Published 2013 by John Wiley & Sons, Ltd.

found in bacteria. They are saprophytic organisms which are widespread in the environment and their principal role in nature is recycling organic matter. They prefer to grow at temperatures below 25° C and not at 37° C. Hence, only a few species are pathogenic to humans; most are opportunist and not obligate parasites. However, the infections they do cause are often persistent and difficult to treat. Since they are so abundant in the environment they frequently cause spoilage of pharmaceuticals and food. Their ability to break down a wide range of organic matter gives them a huge metabolic capability enabling them to synthesize useful products such as foodstuffs, vitamins, organic acids, steroids, enzymes and antibiotics.

4.2.1 Fungal morphology

The morphology of fungi is highly diverse and many species exist in different morphological forms. Most of this is not relevant to us and so for the purposes of this book we will consider fungi as falling into four broad morphological categories:

4.2.1.1   Yeasts

Yeasts are spherical or ovoid unicellular bodies typically 2–4 μm in diameter. Most reproduce by budding, but some like *Schizosaccharomyces rouxii* reproduce by binary fission. Budding is a process of reproduction where the offspring emerge as a bud on the side of the parent cell. This gradually increases in size and eventually pinches off, forming a daughter cell and leaving an area of scarring. A single parent cell can produce up to 24 offspring. A common example is *Saccharomyces cerevisiae* (baker's yeast, brewer's yeast) shown in Figure 4.1, and *Cryptococcus neoformans* is the only significant pathogen which causes the lung infection cryptococcosis.

4.2.1.2   Yeastlike fungi

These often behave like typical budding yeasts but under certain cultural conditions the buds become elongated to form pseudohyphae which are elongated filaments. The most important member of this group is *Candida albicans* (Figure 4.2), which is found as part of the normal microflora of the body. *C. albicans* can cause infections known collectively as candidiasis and involve the mouth, vagina, intestinal tract and lungs.

4.2.1.3   Dimorphic fungi

This morphological grouping contains a number of important human pathogens. As the name suggests these

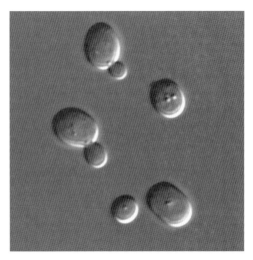

Figure 4.1   Budding *Saccharomyces cerevisiae*. *Source:* http://commons.wikimedia.org/wiki/File:S_cerevisiae_under_DIC_microscopy.jpg.

fungi exist as two distinct morphological forms. They grow as yeasts or filaments depending upon the cultural conditions. At temperatures below 22° C they grow in a filamentous form while at 37° C (body temperature) they grow in the yeast form. *Histoplasma capsulatum* (Figure 4.3) is the most important example and causes a disease known as histoplasmosis. This condition can manifest itself from a mild chest infection through to a fatal disseminated disease.

4.2.1.4   Filamentous fungi

Multicellular moulds grow as long slender branching filaments called hyphae, which are typically 2–10 μm in diameter. These hyphae may be nonseptate (coenocytic) or septate (with cross walls), growing over the surface of a food or medicine substrate and called the

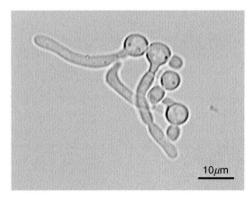

Figure 4.2   Pseudohyphae of *Candida albicans*. *Source:* http://commons.wikimedia.org/wiki/File:C_albicans_germ_tubes.jpg.

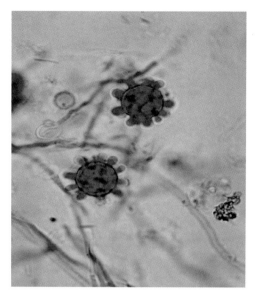

Figure 4.3 *Histoplasma capsulatum. Source:* PHIL ID #4023, Photo Credit: Libero Ajello, Centers for Disease Control and Prevention.

mycelium. Growth occurs by elongation at the hyphal tip. Because of the nature of their multicellular growth we cannot use cell number as a means of quantifying filamentous fungi. Instead it is necessary to monitor growth by measuring mass – for example, dry weight.

Lower fungi like *Mucor hiemalis* (Figure 4.4) and *Rhizopus stolonifer* produce nonseptate hyphae where the cytoplasm is freely diffusible along the filament. Higher fungi, such as *Penicillium* and *Aspergillus,* produce hyphae with septa but in this case the septa do not enclose individual cells. The presence of a pore in each septum allows the cytoplasm and even nuclei to diffuse along the filament. Adjacent filaments can fuse to allow exchange of contents.

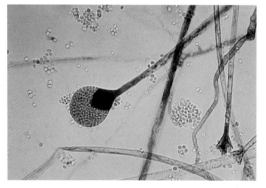

Figure 4.4 Sporangiophore of *Mucor hiemalis. Source:* PHIL ID #3961, Photo Credit: Dr. Lucille K. Georg, Centers for Disease Control and Prevention.

The branching filaments constitute the somatic structure of the fungus and are very similar from one species to another. Under correct environmental conditions the organism will switch from vegetative growth to a reproductive mode. This reproductive mode enables the fungus to propagate the species by moving to new substrates.

4.2.2 Reproduction in fungi

Fungi can exhibit both asexual and sexual reproduction:

4.2.2.1 Asexual reproduction

This is sometimes called somatic or vegetative reproduction. It does not involve the union of nuclei, sex cells or sex organs. In this process all progeny are identical to the parent cells and it is the process primarily responsible for propagating the species.

4.2.2.2 Sexual reproduction

Involves the union of two compatible nuclei and thus is primarily responsible for introducing genetic variation. The majority of fungi produce reproductive organs from a portion of the mycelium while the remainder continues its normal somatic activities.

Both of these processes are extremely diverse across the broad range of fungi and there is little value in discussing the individual reproductive mechanisms of many different species. Sexual reproduction, in particular, occurs less frequently than asexual reproduction and will only be dealt with superficially. However, it is necessary for us to have some grasp of asexual reproduction simply because it is such a common process and because it is the means by which many fungi infect their hosts and contaminate products. In addition, the asexual spore-bearing structures are often the major distinguishing features by which the organism is identified.

There are four common methods of asexual reproduction:

Binary fission

This is comparable to bacterial cell division where the parent cell divides to form two identical daughter cells. It occurs in some species of yeast cells.

Budding

This is the typical method of asexual reproduction for most yeast cells and was described earlier. Note that in the binary fission process the parent cell does not exist after the division has taken place. With budding there is a clear distinction between the parent cell and the daughter cell.

Table 4.1 Asexual spores of medically important fungi.

Asexual spore	Description	Morphology
Aleuriospore	Also called microconidia and macroconidia. Some occur singly or in groups on short lateral branches from the hyphae, others occur directly on hyphae, e.g. the Dermatophytes	
Arthrospore	Arise by fragmentation of the hyphae. Can be highly infectious, e.g. *Coccidioides immitis*	
Chlamydospore	Thick walled spores that form at the end of hyphae or in between hyphal segments. Very resistant to heat and drying, e.g. *Candida albicans*	
Blastospore	Formed by yeasts during budding. Can be formed by spherical yeasts and by elongated pseudohyphae, e.g. *Blastomyces spp.*	
Conidiospore	Occur singly or in groups at the end of specialized structures called conidiophores, e.g. *Penicillium spp.* and *Aspergillus spp.*	
Sporangiospore	Form within sac-like structures known as sporangia found at the end of specialized structures called sporangiophores, e.g. *Mucor spp.* and *Rhizopus spp.*	

Fragmentation

This can sometimes be thought of as a form of spore formation but there are no specialized structures involved (Table 4.1). Here the hyphae simply break up into component segments called arthroconidia which can be dispersed on the wind to other environments. After being formed they can aggregate together in a protective covering in order to survive harsh conditions perhaps imposed by winter weather.

Formation of spores

In this process the younger parts of the mycelium continue to grow across the substrate while in the older parts an abundance of spore-bearing structures arise. These structures are very varied in shape and size but are responsible for producing thousands of spores, each of which is genetically identical to the parent. When they are mature the spores are released and are light enough to be borne on the wind to find fresh food substrates. The most commonly found forms of asexual spores are shown in Table 4.1.

4.3 Commercially important fungi

Some fungi are important because of the damage they cause due to spoilage or because they can be exploited commercially. The following are just a few examples.

4.3.1 *Rhizopus stolonifer* and *Mucor hiemalis*

Both are members of the Zygomycetes (lower fungi) and reproduce asexually by means of sporangiospores produced within sporangia (see Figure 4.4). They are terrestrial saprophytic fungi which are widespread in the environment and are common contaminants. Both are important spoilage organisms and have the capacity to produce an abundance of enzymes. They are used commercially to produce a large number of organic acids – fumaric, lactic, citric and so forth, and are also used in the production of steroids (see Chapter 20).

4.3.2 *Claviceps purpurea*

This fungus is a contaminant of the cereal rye. It is mainly of historical interest from the perspective of infectious disease but the fungus is also an important source of pharmaceutical products. Spores penetrate the developing ears of the rye plant and establish themselves to form a hard resting stage termed a sclerotium (Figure 4.5). This falls to the ground during harvesting, where it overwinters and then germinates the following year, in order to infect the subsequent harvest. The presence of the sclerotium has been responsible for the disease called ergotism which has occurred throughout the last thousand years in central Europe and has killed hundreds of thousands of people.

4.3.2.1 Characteristics of ergots and ergotism

- During milling the ergots (sclerotia) are not separated from grain and the contents become incorporated into flour; this leads to slow poisoning of the population.
- First symptoms are coldness of extremities, followed by sensation of intense burning. Gangrene, necrosis and death may follow.
- Ergots contain a range of alkaloids having a wide spectrum of biological activity:
  ○ controlling haemorrhage;
  ○ induction of childbirth;
  ○ treatment of migraine.
- Now used commercially to produce ergotamine and ergometrine but cannot be cultivated in the laboratory.
- Last major outbreak in France in 1954; 200 people affected; four died.

4.3.3 *Aspergillus niger*

This is a very widely distributed fungus and is abundant in the environment. It is a member of the Deuteromycetes (higher fungi) which is an unusual group in that sexual reproduction has never been observed although the rest of their lifestyle suggests that it should. It is a spoilage organism contaminating crops like hay, nuts and grain. Contamination can lead to production of mycotoxins (aflatoxins), which can cause liver damage if ingested. A number of species of *Aspergillus* cause a disease called aspergillosis. These include a lung infection – 'farmer's lung' and also ear infections. They often grow as a solid mass within body cavities and these are termed aspergillomas.

While still young the mycelium produces an abundance of conidiophores (see Figure 4.6). These are not

Figure 4.5 Ear of rye infected with *Claviceps purpurea* showing a sclerotium. *Source:* R. Altenkamp, Berlin; http://eo.wikipedia .org/wiki/Dosiero:Mutterkorn_090719.jpg.

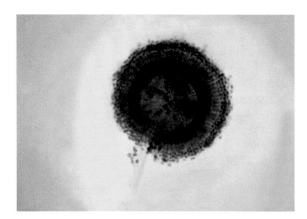

Figure 4.6 *Aspergillus* conidiophore. *Source:* PHIL ID #3965, Photo Credit: Dr. Lucille K. Georg, Centers for Disease Control and Prevention.

organized but arise individually from the somatic hyphae. Conidia are produced in their thousands and are pigmented, giving the colony its characteristic colour (see also Figure 1.6). The fungus has great enzymic activity and is used in the commercial production of a range of enzymes such as amylases, diastase and proteases. It is also used to produce organic acids like citric, gluconic etc. and also fermented products like sake and soy sauce.

4.3.4 *Penicillium chrysogenum*

This fungus is as common in the environment as the *Aspergillus* fungus. It is also a member of the Deuteromycetes (higher fungi) and again no sexual reproduction has been observed. The life history is similar to *Aspergillus* although the morphology of spore-bearing structures is different. The conidiophores have a characteristic brushlike appearance (Figure 4.7) and the colour of conidia is variable.

It is a frequent contaminant of foods and is often seen as a green or blue mould growing on food left around. It is widely used in the production of organic acids such as citric, fumaric, oxalic, gluconic acids and so forth. In the food industry it is also used in the production of veined cheeses like Roquefort, Stilton and Danish Blue. However, it is probably best known for its role in the production of penicillins. The natural product of the fungus is Penicillin G, and from that the whole range of other penicillins such as penicillin V, ampicillin, amoxicillin and others are produced semisynthetically (see Chapter 10).

Figure 4.7 *Penicillium* showing hyphae and developing conidiophores. *Source:* http://commons.wikimedia.org/wiki/File:Penicillium.jpg.

4.4 Pathogenic fungi

It is not the intention to provide an exhaustive list of pathogenic fungi but merely to indicate the types of fungi and the breadth of infections produced. Table 4.2 shows the main pathogenic fungi and the diseases they cause.

As we previously grouped fungi according to their morphological characteristics it is also appropriate to use this scheme as a means of dividing the main pathogens into groups.

- **Yeasts**
  - *Cryptococcus neoformans*
- **Yeastlike fungi**
  - *Candida albicans*
- **Dimorphic fungi**
  - *Histoplasma capsulatum*
  - *Coccidioides immitis*
  - *Blastomyces dermatitidis*
  - *Paracoccidioides braziliensis*
- **Filamentous fungi**
  - Dermatophyte fungi
  - *Aspergillus*
  - *Mucor*

Some of these will be expanded below.

4.4.1 Yeasts

Cryptococcus neoformans is a single-celled yeast possessing a polysaccharide capsule. It reproduces by budding and is the only species of *Cryptococcus* capable of growing at 37° C. The fungus causes cryptococcosis, an infection that is contracted chiefly via inhalation although it can occur through the skin. Initial pulmonary infection may be mild but this can progress to systemic cryptococcosis where the liver, bones and skin may be infected. The most dangerous form is cryptococcal meningitis. The organism is often present in the soil and in the excrement of birds and bats where it can remain viable for months. Hence, poultry workers and cave explorers are potentially at risk.

4.4.2 Yeastlike fungi

Candida albicans is the main organism of interest and is part of the normal flora of the mouth, gut and vagina. Its numbers are kept in check by competition with the other microflora but under some circumstances it can overgrow. This may occur under conditions of antibiotic therapy,

Table 4.2 Some medically important fungi.

Disease type	Example	Microorganism
Systemic mycoses	Cryptococcosis	*Cryptococcus neoformans*
	Coccidioidomycosis	*Coccidioides immitis*
	Histoplasmosis	*Histoplasma capsulatum*
	Blastomycosis	*Blastomyces dermatitidis*
	Bronchpulmonary candidiasis	*Candida albicans*
	Gastrointestinal candidiasis	*Candida albicans*
	Endocarditis	*Candida albicans*
	Pneumonia (PCP)	*Pneumocystis jiroveci (formerly carinii)*
	Aspergillosis	*Aspergillus fumigatus*
	Phycomyosis	*Mucor sp., Rhizopus sp.*
Mucocutaneous mycoses	Oral candidiasis	*Candida albicans*
	Vulvovaginal candidiasis	*Candida albicans*
Subcutaneous mycoses	Sporotrichosis	*Sporothrix schenckii*
	Maduromycosis	*Madurella mycetomi*
	Chromomycosis	*Phialophora verrucosa*
Cutaneous mycoses	Ringworm	*Microsporum sp.*
		Trichophyton sp.
		Epidermophyton floccosum
	Intertriginous candidiasis	*Candida albicans*
Superficial mycoses	Pityriasis versicolor	*Melassezia furfur*
	Tinea nigra	*Exophiala werneckii*
	White piedra	*Trichosporon cutaneum*
	Black piedra	*Piedraia hortai*

diabetes, vitamin deficiency, long-term steroids, immunosuppressive therapy, alcoholism and inappropriate diet.

Candidiasis of the mucous membranes is termed 'thrush' and occurs both in the mouth (see also Figure 9.3) and the vagina. Cutaneous candidiasis is an infection of the skin, while of more concern is bronchopulmonary candidiasis (infection of the bronchi and lungs).

4.4.3 Dimorphic fungi

Histoplasma capsulatum causes a disease called histoplasmosis (Darling's disease), which is endemic in central areas of the United States (Kansas, Ohio) also Africa and the Far East. 95% of cases are subclinical and are only detected by a skin test. In Kansas 80–90% of the population show a positive skin test by the age of 20, while in the United States as a whole 40 million people are

subclinically infected and an estimated 200 000 new cases emerge each year. Of these 5% may go on to develop chronic lung disease while 0.2% may develop disseminated disease. The organism is associated with bird excrement, particularly that of starlings, which pose a threat as they roost in flocks containing millions of birds.

A number of other dimorphic fungi are responsible for similar diseases to histoplasmosis (see Table 4.2) but they will not be dealt with here.

4.4.4 Filamentous fungi

4.4.4.1 Dermatophyte fungi

This is the term given to a range of different fungi which cause diseases of the skin, nails and hair in humans (see Table 4.3). They only affect the keratinized areas of the

Table 4.3 Dermatophyte infections of different sites.

Tinea	Area affected	Alternative name
Capitis	Scalp	Ringworm
Corporis	Body	Ringworm
Cruris	Groin	Jock Itch
Pedis	Feet	Athlete's foot
Barbae	Hair	Barber's itch
Unguium	Nails	Onychomycosis

body and the infections they cause are not life threatening but can be very persistent. The infections are referred to as Tinea infections, also dermatomycoses and ringworm. The name ringworm has arisen because the circular lesions on the skin resemble a worm under the skin (see Figure 4.8). More details on athlete's foot can be found in the text box on the right.

There are three main genera which are of importance:

Epidermophyton
- One of several dermatophytes causing athletes foot.
- Single species – *E. floccosum* attacks skin and nails.
- It is the commonest cause of ringworm of the groin.

Microsporum
- Attacks hair and skin but not nails.
- Infects skin of children – causes tinea capitis.
- May be spread by handling cats and dogs.

Trichophyton
- Attacks skin, nails and hair. *T. mentagrophytes* var interdigitale is the main cause of athlete's foot.

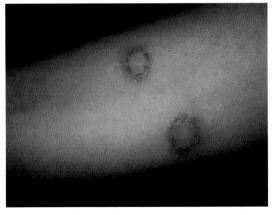

Figure 4.8 Ringworm lesions on the forearm. *Source:* http://commons.wikimedia.org/wiki/File:Herpes_circin%C3%A9_01.jpg.

Athlete's foot

- It is one of the most common infectious diseases.
- It affects males and females of every age and race.
- It is highly contagious; it may be spread by contaminated floors and towels.
- A single spore can initiate infection.
- The spore germinates and penetrates the stratum corneum.
- The fungus produces branching septate hyphae and forms arthrospores by fragmentation of hyphae.
- The infection begins as cracking or scaling between toes.
- Redness and itching.
- The condition can become chronic especially in humid conditions.
- The spores infect shoes and socks and are not readily killed.

4.4.4.2 *Aspergillus niger*

This organism causes systemic infections that occur more commonly in immunocompromised patients but they can arise in immunocompetent patients where there has been recent tissue damage. The main infections of interest include:

- Invasive aspergillosis.
  - Mainly affects immunocompromised.
  - Lungs affected in 80–90% of patients.
  - Dry cough, fever, chest pain and dyspnoea (shortness of breath).
- Aspergillus tracheobronchitis.
  - Mainly affects AIDS and lung transplant patients.
- Aspergillus sinusitis.
  - More common in bone marrow transplant patients.
- Cerebral aspergillosis.
  - Occurs in 10–20% of patients with invasive aspergillosis.
  - Usually only in immunocompromised.
- Aspergilloma.
  - Can colonize cavities within lungs or sinuses.
  - Ball of fungus develops in the cavity. May be no mucosal involvement. Surgical removal may be curative.
  - Patients may be asymptomatic, but most have persistent productive cough and weight loss.

4.4.4.3 Mucormycosis

This term covers a variety of infections caused by the Zygomycetes, for example *Mucor* and *Rhizopus*.

Spores enter the body via the respiratory tract but can be inoculated through the skin. Generally the diseases are limited to immunocompromised patients or those with diabetes, trauma or solid organ transplants. The infecting fungi are able to grow rapidly and produce abundant spores; hence disease spread can be rapid and is often fatal.

- Rhinocerebral mucormycosis is the most common form.
- Pulmonary mucormycosis is seen mostly in neutropenic patients.
- Cutaneous mucormycosis presents as a chronic non-healing ulcer with central necrosis – it often follows an insect bite or a gardening injury.

Acknowledgement

Chapter title image: PHIL ID #3053, Dr. Arvind A. Padhye, Centers for Disease Control and Prevention.

Chapter 5
Protozoa

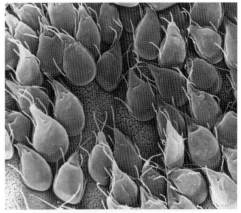

The protozoan *Giardia* pictured here attached
to the lining of the intestine

KEY FACTS

- Protozoa are only of pharmaceutical interest because they cause disease; they are not product contaminants or spoilage organisms in medicines.
- Protozoal infections are much more common in hot countries. Although UK cases of malaria are always caught abroad, there are several endemic pathogenic protozoa, of which *Trichomonas vaginalis*, *Giardia lamblia* and *Cryptosporidium* species are the most important.
- Metronidazole is normally effective for the treatment of *Trichomonas* and *Giardia* infections, but the treatment of some other protozoal infections can be difficult.
- Malaria is caused by five species of *Plasmodium*, of which *P. falciparum* is the most dangerous; *P. vivax*, *P. ovale*, *P. malariae* and *P. knowlesi* usually cause less severe symptoms, and infections by these species are referred to as benign malaria.
- The choice of drug for the treatment or prevention of malaria will be influenced by local resistance patterns and the species of *Plasmodium* responsible.

Protozoa are single-celled eukaryotic organisms that can cause a variety of severe human infections, and they are of pharmaceutical interest for this reason alone. They are extremely unlikely to arise as contaminants of pharmaceutical raw materials, and they are not used in the manufacture of any medicines other than the vaccines that prevent the diseases they cause. As with bacteria, there are very many species of protozoa (some estimates exceed 50 000) but only a relatively small number are capable of causing infections – at least in humans. A few of these pathogenic species are able to survive and reproduce outside the body, for example the pathogenic amoebae, but most are parasites that can only reproduce inside an animal host; several of these, like the organisms causing malaria, leishmaniasis and trypanosomiasis (sleeping sickness), for example, require animal vectors – insects – to transport them from one human victim to the next.

5.1 Cultivation of protozoa

The pathogenic protozoa differ significantly in terms of their shape and size (Figure 5.1). They can be divided into three groups in terms of the way they are grown:

Essential Microbiology for Pharmacy and Pharmaceutical Science, First Edition. Geoffrey Hanlon and Norman Hodges.
© 2013 John Wiley & Sons, Ltd. Published 2013 by John Wiley & Sons, Ltd.

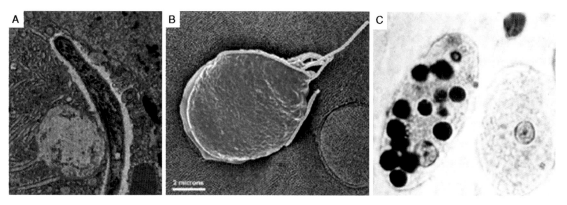

Figure 5.1 Pathogenic protozoa: A. malaria (*Plasmodium*) sporozoite in intestinal epithelium; B. *Giardia* cell attached to an epithelial surface; C. *Entamoeba histolyticum* (the light blue cell) which has ingested host red blood cells (dark circles). A and B are artificially coloured electron micrographs with higher magnification than C. thus, the *Entamoeba* cells are approximately twice the size of the other two. *Sources:* A. http://commons.wikimedia.org/wiki/File:Malaria.jpg. B. PHIL ID #11653; Photo Credit: Dr. Stan Erlandsen, Centers for Disease Control and Prevention. C. CDC DPDx Image Library; http://www.dpd.cdc.gov/dpdx/html/ImageLibrary/Amebiasis_il.htm.

- Amoebae can be cultivated easily in the laboratory in much the same way as bacteria, for instance *Acanthamoeba* species.
- Some can be grown in mixed cultures with companion organisms and, with more difficulty, on their own, for example the pathogenic protozoa possessing flagella, such as *Giardia* and *Trichomonas*.
- Others are extremely difficult to grow in the lab because they have both human and insect stages in their life cycles so they require very complex media supplemented with red blood cells, e.g. the *Plasmodium* species responsible for malaria.

Laboratory cultivation of protozoa is not, therefore, undertaken even in hospital pathology laboratories let alone pharmaceutical quality assurance labs, so this aspect will not be considered further. Diagnosis of protozoal infections does not normally require cultivation of the organism anyway; it is usually based upon symptoms and immunological tests or, in some cases, on microscopic examination of faecal samples to detect the organism's cysts.

5.2 Protozoal infections: the global and UK perspectives

Protozoal infections are far more common in tropical countries than in temperate ones, largely because higher tropical temperatures promote the reproduction of both

the protozoa themselves and their insect vectors. In temperate countries winter frosts tend to kill protozoa, but despite this there are several indigenous protozoal infections in the UK. Only about 10% of the cases of giardiasis, 5% of amoebic dysentery, and 3% of cryptosporidiosis recorded annually by the Heath Protection Agency for England and Wales are associated with foreign travel, so the great majority of these infections are caught within the country. By contrast, all UK cases of malaria and the majority of leishmaniasis infections are caught abroad. This situation may change however, because the geographical ranges of protozoal infections change over the years. Malaria, for example, was relatively common in southern Europe and the southern United States a century ago, and the last recorded indigenous cases in Britain were in the 1950s. The possibility exists that the global ranges of malaria and other tropical protozoal (and some nonprotozoal) infections may expand due to climate change.

Statistics relating to infections may be confusing because of the way they are recorded. Cases (the number of persons with the disease) differ from outbreaks (in which a single source affects many persons), and some databases record the number of new cases per year, which is not the same as the number of sufferers (particularly for diseases of long duration, which may extend over more than one year). Nevertheless, it is clear that the numbers of UK cases recorded in Table 5.1 are minute compared with those globally. Each year, there are approximately 350–500 million cases of malaria worldwide, for example, and more than 66 million people suffer from African trypanosomiasis.

Table 5.1 Incidence of protozoal infections in the UK.

Infection	Protozoan	UK cases per year (approx)
Trichomoniasis	*Trichomonas vaginalis*	6000
Cryptosporidiosis	*Cryptosporidium parvum* and *Cryptosporidium hominis*	4500
Giardiasis	*Giardia lamblia*	3500
Malaria	*Plasmodium falciparum* (approx 75%) and other species	2000*
Toxoplasmosis	*Toxoplasma gondii*	500
Amoebic dysentery	*Entamoeba histolyticum*	100
Acanthamoebiasis	At least 10 species of *Acanthamoeba*	50–100
Leishmaniasis	21 different species of *Leishmania*	60*
Trypanosomiasis	*Trypanosoma brucei* causes African sleeping sickness	1
	Trypanosoma cruzi causes Chagas disease in Central and S America	<1

*Almost all caught abroad

5.3 The characteristics and transmission of the major UK protozoal infections

5.3.1 Trichomoniasis

- *Trichomonas vaginalis* causes the greatest number of UK protozoal infections each year and it is the most common protozoal pathogen in industrialized countries, being estimated to affect over 170 million people globally.
- It is a sexually transmitted disease that infects males and females at approximately the same frequency, but infections in both sexes, though particularly in males, are often without symptoms.
- It commonly invades the vagina, particularly in circumstances when the protective acidity of the vaginal secretions is reduced, but it can also infect the urinary tract, fallopian tubes and pelvis in females, and the prostate in males.
- It is not likely to be life threatening and, although it may spontaneously clear, treatment with metronidazole or tinidazole is normally prescribed (Table 5.2) and achieves a high success rate. Treatment is usually necessary for the patient's sexual partners too, even if they are asymptomatic.
- Chronic untreated infection in females may cause premature births during pregnancy and may increase susceptibility to both HIV and cervical cancer.

5.3.2 Cryptosporidiosis

- The incidence has been rising in recent years so that it is the second most frequent protozoal infection in the UK.
- It causes diarrhoea, abdominal pain and nausea but is usually self-limiting in immunocompetent persons and drug treatment is unnecessary (except for rehydration therapy).
- In developed countries infection without symptoms is rare (approximately 1% of cases), but asymptomatic infection in developing countries is far more common.
- It is a complication in AIDS but the introduction of highly active antiretroviral therapy has reduced the incidence.
- The organism is an obligate intracellular parasite (can only reproduce inside a host cell) but it is capable of producing oocysts that are extremely robust and can survive outside the body for long periods; they even survive water chlorination.
- Transmission is via the faecal-oral route and infection is readily initiated by the low infective dose of 10–1000 oocysts.
- Treatment options in immunocompromised patients are few, and the treatments that have been proposed rarely eradicate the infection without improvement in immune function. Nitoxanide is available in the United States, but only on a named patient basis in the United Kingdom; alternatives mainly comprise the macrolide antibiotics: clarithromycin and azithromycin.

Table 5.2 Drug treatment of protozoal infections.

Infection	Drug therapy
Trichomoniasis	Metronidazole or tinidazole
Cryptosporidiosis	In immunocompetent patients symptomatic treatment (fluid rehydration therapy and anti-diarrhoeal drugs) is normally adequate (nitoxanide is used as an antiprotozoal agent in the United States). No consistently effective antimicrobial agents for immunocompromised patients
Giardiasis	Metronidazole or tinidazole or mepacrine
Malaria	See Table 5.3
Toxoplasmosis	Not usually necessary, but if required pyrimethamine and sulfadiazine administered for several weeks under expert supervision
Amoebic dysentery	Metronidazole or tinidazole followed by a 10-day course of diloxanide furoate
Acanthamoebiasis	Propamidine isethionate
Leishmaniasis	Sodium stibogluconate or amphotericin
Trypanosomiasis	Pentamidine and other drugs under expert supervision

5.3.3 Giardiasis

- *Giardia lamblia* is the only other pathogen causing a significant number of indigenous UK infections; it is transmitted via the faecal-oral route following consumption of contaminated food or water.
- Its cysts resist chlorination, survive for long periods outside the body and, following ingestion, cause an infection resulting in acute diarrhoea.
- The infection is quite frequently asymptomatic, and relatively easy to treat with metronidazole or tinidazole which are the drugs of choice (Table 5.2), or, less commonly, mepacrine.

5.3.4 Other protozoal infections

The other protozoal infections listed in Table 5.1 are, with the exception of malaria (considered in the next section), relatively uncommon in the United Kingdom, but when infections do arise the drug treatments described in Table 5.2 are normally effective.

- Amoebic dysentery although less common than giardiasis and cryptosporidiosis, may cause a more severe dysentery (defined as diarrhoea with blood or mucus in the faeces), which, if untreated, may lead to a fatal liver abscess. This organism, too, is transmitted by the faecal-oral route and produces cysts, although they are less robust than those of *Giardia* and

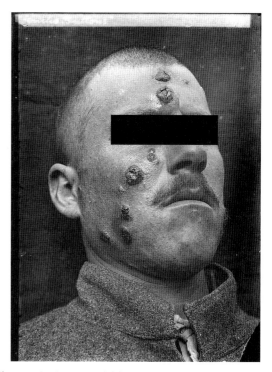

Figure 5.2 Cutaneous leishmaniasis, also known as kala azar, dum-dum fever or 'Jericho buttons'; the last name is relevant here as this photograph, taken in 1917, shows the disease in a soldier serving near the city of Jericho in the Middle East. *Source:* G. Eric, Edith Matson Photograph Collection, US Library of Congress; http://commons.wikimedia.org/wiki/File:Jericho-Buttons.jpg.

Cryptosporidium species and only survive for a few months even in moist conditions. *Acanthamoeba* species are relatively rare problem organisms contaminating contact-lens solutions and have caused ophthalmic keratitis.

- *Toxoplasma gondii*, a protozoan transmitted via cat faeces can, particularly in immunocompromised individuals, cause severe, even fatal, brain damage.
- Trypanosomiasis is, like malaria, transmitted via insect bites, and infects the bloodstream as well as muscle and the central nervous system.
- Cutaneous leishmaniasis, also transmitted via insect bites, is one of the few protozoal infections that is visually recognizable from the skin lesions it causes (Figure 5.2), although in its other form, visceral leishmaniasis, it is much more dangerous because internal organs are infected.

5.4 The transmission, prophylaxis and treatment of malaria

There are about 2000 malaria cases in the UK each year and 10–20 deaths, but the number of deaths worldwide is approximately 1.5 million per year, of which 90% occur in Africa. The relatively small number of cases might give the impression that the disease is of only minor importance from a UK pharmaceutical perspective, but the scale of modern international travel means that antimalarial drugs are dispensed far more frequently for disease prevention than for treatment. Data from the National Health Service Prescription Pricing Division shows that 5.28 million prescriptions were written for antimalarial

Table 5.3 Summary of British National Formulary (BNF) recommendations for the treatment and prophylaxis of malaria.

Antimalarial	Falciparum malaria treatment	Benign malaria treatment	Prophylaxis
Artemether with lumefantrine[a]	Yes	Yes	X
Chloroquine[a]	No longer recommended due to resistance	Yes	Yes, in areas with low risk of chloroquine-resistant falciparum malaria
Mefloquine[a]	Rarely used due to resistance	Rarely used	Yes, in areas with high risk of chloroquine-resistant falciparum malaria
Primaquine	X	Yes, for *P. vivax* and *P. ovale*; used with quinine or chloroquine	X
Proguanil	No	No	Yes, usually with chloroquine
Proguanil with atovaquone[a]	Yes	Yes	Yes, particularly in areas with high risk of resistant falciparum malaria
Pyrimethamine[a]	Not used alone for malaria; only for toxoplasmosis		
Pyrimethamine with sulfadoxine[a]	Possibly, together with quinine	X	Not recommended in the United Kingdom
Quinine[a]	Yes, with doxycycline (adults) or clindamycin (children)	X	No
Doxycycline[a]	Yes, together with quinine	X	Yes, in areas of mefloquine or chloroquine resistance

[a]= prescription only medicine in the UK; X = no BNF recommendations.

drugs in the period from January 2007 to June 2008 and to this number must be added the private prescriptions and the sales of those products which are not prescription-only medicines (Table 5.3). Advice is also sought in pharmacies regarding the suitability of alternative insect repellents, which, together with appropriate clothing and mosquito nets, are an integral part of a protection strategy designed to avoid being bitten by the female *Anopheles* mosquitoes that transmit the infection.

Swamp drainage, the use of insecticides, and biological control agents like predatory fish that eat insect larvae are further strategies used to control mosquitoes and reduce the incidence of the diseases they transmit.

It is important to recognize the difference between the various forms of malaria caused by different species of the *Plasmodium* parasite. *Plasmodium falciparum* is the most common and the most dangerous species, causing what is sometimes referred to as malignant malaria; it has the highest mortality and complication rates and accounts for more than 90% of malaria infections and deaths worldwide (approximately 75% of cases in the United Kingdom).

Four other species of *Plasmodium* may also cause the disease: *P. vivax*, *P. ovale*, *P. malariae* and *P. knowlesi*. They give rise to a less severe infection referred to as benign malaria, which is relatively more common in parts of Asia and South America. The five species exhibit differing susceptibilities to the common antimalarial drugs, so those used to treat or to prevent falciparum malaria are not necessarily recommended for benign malaria and vice versa (Table 5.3). Chloroquine, for example, was formerly far more widely used than it is now, but resistance has become so widespread that it is currently employed largely for the treatment of benign malaria and for prophylaxis in regions with a low risk of resistance in falciparum malaria. The situation is further complicated by variations in malaria risk during different months of the year since rainfall affects mosquito breeding.

Mosquitoes are also less common at high altitude, so people living in mountainous areas of a malaria-prone country might be at little or no risk. Because of the regional variations in susceptibility then prior to travel it is important to consult the specific recommendations relevant for each country or region, such as the guidelines in the British National Formulary or those of the UK Health Protection Agency (http://www.hpa.org.uk/infections/topics_az/malaria/guidelines.htm).

There have been many attempts to produce a malaria vaccine, but the complexity of the organism's life cycle and its rapid rate of reproduction and mutation have posed major problems. A recent candidate vaccine has been trialled in Mozambique and the Gambia and was reported to be 71% effective in providing short-term protection against falciparum malaria.

Acknowledgement

Chapter title image: PHIL ID #11632, Photo Credit: Dr. Stan Erlandsen, Centers for Disease Control and Prevention.

Chapter 6
Viruses and viral infections

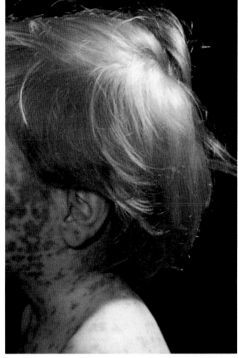

Rubella, otherwise known as German measles

KEY FACTS

- Viruses do not have a cellular structure and the simplest of them consist merely of nucleic acid surrounded by protein.
- All cellular organisms including animals, plants, bacteria, fungi and other microorganisms are vulnerable to virus infections.
- Viruses cannot be grown in simple media on Petri dishes. They can only be grown in suitable host cells.
- They can survive but cannot reproduce in medicines, so they are not potential spoilage organisms.
- Some viruses possess a lipid envelope; these are normally more susceptible to drying, solvents and disinfectants than nonenveloped ones.
- Viruses are not susceptible to common antibacterial antibiotics.
- Most viruses have a fairly limited range of hosts and they are transmitted from one individual to another by a wide variety of mechanisms.
- Some viruses may be latent (dormant) and survive in their host cells without killing them. They may be activated at a later date to reproduce and kill the host cell.
- Some viruses cause cancer.
- Most human viruses survive outside the body only for short periods and most of them are at least as vulnerable to heat, extremes of pH, drying and disinfectants as bacteria.

Essential Microbiology for Pharmacy and Pharmaceutical Science, First Edition. Geoffrey Hanlon and Norman Hodges.
© 2013 John Wiley & Sons, Ltd. Published 2013 by John Wiley & Sons, Ltd.

6.1 The pharmaceutical importance of viruses

Viruses are the most abundant organisms on the planet and their numbers exceed the global human population by a factor of 10^{21}, or, putting it another way, for every human being on the planet, there are 1 000 000 000 000 000 000 000 viruses. They are parasites that can only reproduce inside a host cell; they have no metabolism of their own and possess few, if any, enzymes. All cellular organisms are vulnerable to virus infection; not just animals and plants, but bacteria and all other kinds of microorganisms too. Viruses do not, themselves, have a cellular structure and the simplest of them consist merely of nucleic acid surrounded by protein, so it is even debateable whether or not viruses should be regarded as living organisms. Just as the entire human genome has been mapped in recent years, so too have the genomes of several viruses, and in 2002 the polio virus was the first to be totally synthesized in the laboratory. Because they do not have a cellular structure, viruses do not possess mitochondria, ribosomes or the other organelles that are required for normal cell function and metabolism, and this factor further supports the view that they should be regarded as self-assembling clusters of complex chemicals rather than living organisms.

Viruses cannot be cultured on simple media in Petri dishes and they certainly cannot grow in medicines so, unlike bacteria, they have no potential as spoilage organisms. They can only be grown in an appropriate host cell; this means that a laboratory equipped for the cultivation of fertilized chickens' eggs or mammalian cells must be available for viruses to be studied. Such laboratories are expensive to operate and require skilled personnel so, despite the fact that viruses might survive in a medicine without actually growing in it, tests to detect or count contaminating viruses are rarely undertaken. Fortunately, most of them only survive for a short time outside the host organism (although there are important exceptions, such as hepatitis viruses), so medicines do not normally represent a major source of viral infection.

Viruses are totally resistant to the commonly used antibiotics, so they are of pharmaceutical interest not only because of the severity of the infections they may cause but also because of the difficulties of treating such infections. There are, however, two areas where viruses have applications in medicine, although it must be stressed that these are still areas of research and development rather than of widespread established use. They have potential as drug delivery systems, particularly for diseases like cystic fibrosis that are a consequence of genetic disorders and for which 'gene therapy' might be a realistic option, because viruses may act as a carrier by which 'normal' genes may be introduced into affected human cells and tissues. The second application is for viruses that infect bacteria (bacteriophages, or 'phages' for short) to be used as alternatives to antibiotics for the treatment of human infections. Phage therapy, as it is called, was of more widespread interest in the first half of the twentieth century, but this interest lapsed in Western Europe and North America with the advent of the antibiotic era. The current problem of increasing antibiotic resistance has reawakened interest in phage therapy, but a phage-based medicine is still some way from receiving licensing approval for Western markets.

6.2 Virus structure and replication

Although viruses are not the smallest or simplest infectious agent known (viroids, consisting merely of RNA without surrounding protein, and prions, which are self-replicating proteins without nucleic acid, are both smaller and simpler) viruses are the smallest that can readily be seen with an electron microscope. Figure 6.1 shows electron micrographs of six different viruses selected to illustrate their variety of shapes and sizes. The individual virus particles (called virions) typically range from about 20 nm in diameter (Figure 6.1A), which is about one-fiftieth of the size of a typical bacterium like *Staphylococcus aureus*, up to about 350 nm which is the size of the so-called pox viruses (for example, the herpes group and the viruses responsible for chicken pox, shingles and smallpox (Figure 6.1F).

Many, particularly the smaller ones, appear to be approximately spherical (Figures 6.1A and 6.1B), but higher resolution electron microscope images show that the structure is frequently not a true sphere but an icosahedron (comprising 20 triangular sides; Figure 6.2). The protein coat that surrounds the nucleic acid and protects it from mechanical and chemical damage is termed the capsid and is made up of individual units called capsomeres; the nucleic acid and surrounding capsid are together termed the nucleocapsid. The common alternative to an icosahedron is for the capsid to have a helical structure, which, on an electron microscope, appears simply as a straight rod (Figure 6.1C); more complex structures also exist including those with tails, which are common amongst

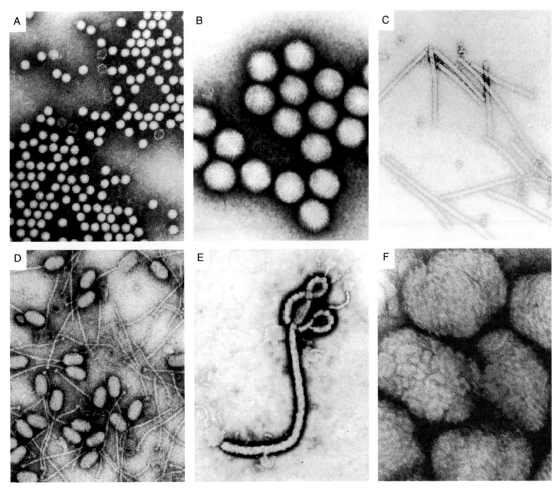

Figure 6.1   Electron microscope images of viruses illustrating differences in shape and size. As displayed, the photographs reflect their true relative size (in brackets below). A. Polio virus (25–30 nm). B. Adenovirus (causing human respiratory infections, 75–100 nm). C. Tobacco mosaic virus (300 × 20 nm). D. Bacteriophage 3A of *Staphylococcus aureus* (300 nm including the tail). E. Ebola virus (800–1000 nm long). F. Smallpox virus (350 × 250 nm). *Sources:* A. PHIL ID #235; Photo Credit: Dr. Joseph J. Esposito and F.A. Murphy, Centers for Disease Control and Prevention. B. PHIL ID #237; Photo Credit: Dr. G. William Gary, Jr., Centers for Disease Control and Prevention. C. http://commons.wikimedia.org/wiki/File:TobaccoMosaicVirus.jpg. D. http://commons.wiki-media.org/wiki/File:Phage_de_S_aureus_3A.jpg. E. PHIL ID #1181; Photo Credit: Frederick A. Murphy, Centers for Disease Control and Prevention. F. PHIL ID #2292; Photo Credit: Frederick A. Murphy, Centers for Disease Control and Prevention.

bacterial viruses (Figure 6.1D), and long curved or coiled filaments (Figure 6.1E).

When a virus infects a human cell it may cause immediate death and lysis of that cell, or it may cause a persistent infection whereby the cell is not immediately killed but releases (sheds) new virus particles steadily over a period of time. In order to persistently shed new virions however, the virus must avoid damaging the host cell membrane and it does this by allowing the newly created virions to 'bud' from the host cell (Figure 6.3).

Budding involves the new virus particle becoming wrapped in host cell membrane, which, on its outer surface, has viral receptor glycoproteins that enable the new virion to attach to a new host cell; an envelope without such receptors would much reduce its infective capacity. Viruses that have adopted this shredding strategy therefore, are said to be 'enveloped' because they have a phospholipid bilayer membrane around them. The envelope does not afford any significant degree of protection against physical or chemical damage and is, itself, easily removed or damaged by surfactants and lipid solvents. So, although it might be expected that an additional layer outside the capsid would make the virus particle more robust, in fact the opposite is the case and

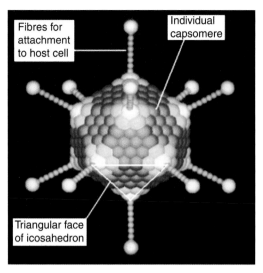

Figure 6.2   Computer-generated diagram of adenovirus showing its icosahedral structure. *Source:* Adapted from an image by Dr. Richard Feldmann, National Cancer Institute; http://commons.wikimedia.org/wiki/File:Adenovirus.jpg.

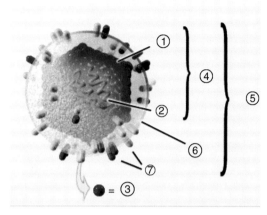

Figure 6.4   Structural components of a generalized virus particle: 1. capsid; 2. nucleic acid; 3. individual capsomeres; 4. nucleocapsid, 5. virion; 6. envelope; 7. spike glycoproteins. *Source:* http://commons.wikimedia.org/wiki/File:Virion.png.

enveloped viruses, such as herpes, measles, mumps, rubella, chicken pox, influenza and HIV, tend to be more sensitive to disinfectants and lipid solvents than nonenveloped ones, such as polio, adenovirus, rhino virus (common cold) and hepatitis A (see below).

Viruses normally contain only one type of nucleic acid, either DNA or RNA, which may be single or double stranded and is tightly coiled to fit into the available space within the capsid; it typically codes for between 10 and 200 genes depending on the size of the virus. Retroviruses, including HIV, are exceptional in that they contain both nucleic acids. The type of nucleic acid, capsid symmetry (icosahedral, helical or complex), the possession of an envelope or not, and the nucleic acid structure (single or double stranded) are the criteria by which viruses are classified. In addition to the structural (capsid) protein, viruses possess protein in the form of enzymes (exceptionally up to ten different types, but usually fewer, or none at all), which are commonly but not exclusively associated with reproduction of the viral genome. Thus the most complex virus is likely to exhibit a structure similar to that shown in Figure 6.4 and contain nucleic acid, protein, phospholipid and glycoproteins (sometimes referred to spike proteins).

6.3 Viral infections

Viruses usually have quite a limited host range. Rabies virus is rather unusual in that it can infect a variety of mammals; smallpox on the other hand has just one host – humans. Most mammalian viruses are somewhere between these two extremes and are able to infect a small number of related species – for example HIV, which infects humans and some other primates. Many bacteriophages are even more specific, attacking only a few strains within a species.

A virus must have an effective mode of transmission from one person to another. If it regularly killed its human host before being transmitted to another individual it would have a reduced chance of survival and would risk extinction. Transmission is often described as being either vertical (meaning through generations – from pregnant mother to embryo or from mother to baby via breast milk), or horizontal, where the virus is transmitted from one individual to another of the same species and they are not in a parent-child relationship.

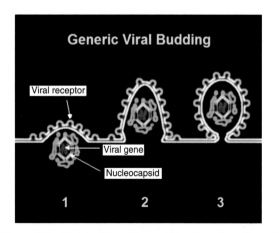

Figure 6.3   Viral budding. *Source*: http://en.wikipedia.org/wiki/File:Budding_of_generic_virus,_pictorial_represent.jpg.

Several modes of transmission occur and these, not surprisingly, are related to the principal site of the virus infection in the body.

- Via droplet nuclei (particles expelled into the atmosphere during sneezing, coughing or talking), for example influenza, common cold, measles and other viruses infecting the respiratory system.
- The faecal-oral route is the common means of transmission for viruses whose *primary* infection site is the gastro-intestinal tract or associated organs, for example hepatitis A and polio.
- Via sexual intercourse, for instance HIV/AIDS, hepatitis B, genital herpes and cervical cancer.
- Insect vectors transmit dengue fever virus, West Nile disease and tickborne encephalitis.
- Direct contact with infected patients or contaminated objects by which the virus is introduced onto the skin, for example warts and verrucae and, in some cases, then into the blood stream by skin damage following scratching, for instance pox viruses.
- By direct introduction into the blood stream, for example hepatitis B, contaminating addicts' syringes and needles, and rabies following animal bites.

The ease by which an infection is transmitted via an aerosol depends upon humidity, because the virus is initially expelled from the infected lung as a mucus-coated droplet, but in low humidity (e.g. centrally heated houses in winter) the mucus rapidly dries so the aerosolized particle becomes lighter and remains suspended in the air for longer; it is partly for this reason that the common cold and influenza are more widespread in winter. Because viral envelopes are susceptible to damage by heat and drying the viruses that possess them normally survive outside the body only for short periods, typically a few hours, so they are more reliant on rapid person-to-person transmission via aerosols or direct contact.

In order to replicate, a virus needs to be able to recognize and attach to a host cell, then to get inside; merely being adsorbed onto the outer surface of the cell is not enough. Recognition is achieved via the glycoproteins, which are the spikes of enveloped viruses or components of the capsid of nonenveloped viruses (Figure 6.2); these attach to receptor molecules on the host. Penetration is achieved via a process which may be considered the opposite of budding, whereby the attached virus is engulfed by the host plasma membrane (Figure 6.5).

After entering the cell, the infecting virus is transported to the area where the new virions will be assembled: RNA viruses remain in the cytoplasm, whereas DNA viruses go to the nucleus. The capsid is

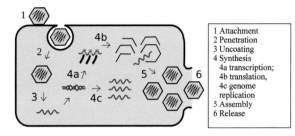

Figure 6.5 Virus replication cycle. *Source:* Wikimedia Commons: http://commons.wikimedia.org/wiki/File:Virus_Replication_Cycle.svg.

removed (termed 'uncoating') to expose the nucleic acids to the enzymes responsible for transcribing the viral genes. This is followed by assembly of the new virus particles and their release by budding or cell lysis. The time required for the complete replication cycle depends upon the environmental conditions and the generation time of the host cell. In rapidly growing bacterial cultures infection with a phage particle can lead to the creation and release by cell lysis of approximately 200 new virions (this is known as the 'burst size') within 20 minutes, but the process is much slower in mammalian cells and the burst size more variable; in some cases thousands or tens of thousands of new viruses are released. The liberated particles infect adjacent cells, and if those cells are themselves attached to, and covering, a surface (either a biological membrane as in a fertilized chicken's egg or the plastic surface of a laboratory culture vessel for

Figure 6.6 Virus plaque assay. Kidney cells were grown to completely cover the bottom of wells in a plastic tray. The wells were infected with successive fourfold dilutions of herpes simplex virus and the cells cultured overnight to allow the formation of plaques (so the well at the top left received 4 times the virus concentration of the well immediately to its right, which, in turn, received 4 times the concentration of the third one in the row, and so on). The living cells were stained blue, but the plaques containing no living cells (seen most clearly in the two wells on the top right) show up as small colourless circular zones in the otherwise uniform blue cell monolayer. *Source:* http://commons.wikimedia.org/wiki/File:Plaque_assay_dilution_series.jpg.

example), the successive replication cycles produce a circular zone of cell lysis of several mm diameter called a plaque. If it is assumed that each plaque is formed from a single virus particle or an aggregate of several particles (either being described as a plaque forming unit – abbreviated to PFU) their formation and numbers form the basis for a method of counting viruses. This is called a 'plaque assay' and it would be required, for example, during the manufacture of a viral vaccine (Figure 6.6).

6.3.1 Latent viral infections

When a virus infects a human cell it usually reproduces and causes the cell to die and lyse in order to release the new virus particles, but that does not always happen. Some viruses may enter a latent (dormant) state in a small fraction of the infected cells; they do not cause immediate damage but give rise to a persistent infection. The dormant virus may exist free in the cytoplasm or become incorporated into the host cell's DNA and remain there for long periods – possibly throughout the life of the person concerned. In either case, suitable stimuli can reactivate the latent virus and so cause it to reproduce and kill the cell in the 'normal' manner. In humans, this reactivation may occur days, months or many years after the initial infection.

This situation is particularly common amongst herpes viruses, so cold sores due to herpes simplex (Figure 6.7) may recur following the reactivating stimulus of exposure to the ultraviolet component of sunlight or another viral infection (typically a common cold – hence the name), and the varicella zoster virus, which causes chicken pox in childhood, may be reactivated to cause shingles in adult life. Retroviruses, including HIV, become integrated into the host DNA and are almost

impossible to remove without killing the cell. The phenomenon of viral latency also occurs amongst bacteriophages and is a mechanism by which antibiotic resistance genes are transferred from one bacterial cell to another by means of a 'phage vector (see the passage on transduction in Chapter 13).

6.3.2 Viral cancers

In addition to causing some of the most severe infections with the highest mortality rates and being implicated in the spread of antibiotic resistance, viruses have another major impact upon human health in that they can initiate several forms of cancer. It has been estimated that in 2002 approximately 1 in 6 human cancers worldwide were of viral origin, with the human papilloma viruses being responsible for approximately 5% of the total (mostly cases of cervical cancer) and the hepatitis B and C viruses together being responsible for a further 5% (causing liver cancer). Some herpes viruses (Figure 6.8), HIV and the human T-cell leukaemia viruses are also associated with cancer, though significantly less frequently than papilloma and hepatitis viruses. The term 'oncogenic' is sometimes used to describe viruses with cancer-causing ability and these have been the subject of intensive research in recent years because of the possibility of creating vaccines that would protect susceptible individuals against the forms of cancer in question. The first fruits of that research have been the two forms of cervical cancer vaccine first marketed worldwide in 2006.

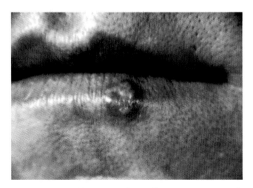

Figure 6.7   A 'cold'sore' resulting from a persistent herpes simplex infection. *Source:*   PHIL ID #1573; Photo Credit: Dr. Herrmann, Centers for Disease Control and Prevention.

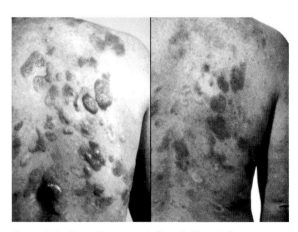

Figure 6.8   Kaposi's sarcoma before (left) and after treatment with interferon. This cancer is caused by human herpes virus 8. *Source:* National Cancer Institute; http://commons.wikimedia.org/wiki/File:Kaposi%27s_sarcoma_before.jpg (left). National Cancer Institute; http://commons.wikimedia.org/wiki/File:Kaposi%27s_sarcoma_before.jpg (right).

Table 6.1 Activity and required contact time for disinfectants acting on enveloped and nonenveloped viruses.[a]

Active ingredient	Enveloped viruses[b]	Nonenveloped viruses[c]	Contact time required (min)
Ethanol-based 60–95%	Good	Poor	Rapid 0.5–2.0
Hypochlorite (4–6% chlorine)	Good	Good	Rapid 0.5–2.0
Iodine-based 0.5–5%	Good	Fair	Medium 5–10
Phenols 0.2–3%	Fair	Poor	Medium 5–10
Quaternary ammonium compounds 2%	Good	Poor	Medium 5–10
Hydrogen peroxide 3% or less	Fair	Poor	Slow 10–20

[a]Based on New York State's categorization of disinfectants for hospital use.
[b]Including herpes, simplex, HIV, hepatitis C, cytomegalovirus, measles, mumps, rubella, influenza, respiratory syncitial virus, varicella zoster, coronavirus and hepatitis B, which, although not strictly enveloped, has similar sensitivity.
[c]Including hepatitis A, coxsackie, polio, rhinovirus, human papilloma virus, adenovirus, rotavirus and parvovirus.

6.4 Virus survival outside the body and susceptibility to disinfection

Apart from their possible roles as a drug-delivery system and in phage therapy (mentioned above) the main pharmaceutical interest in viruses is killing them, or at least preventing or treating the infections they cause. An understanding of their potential to survive outside the body and their susceptibility to physical and chemical methods of inactivation is clearly relevant in pharmaceutical science.

The environmental factors that influence a virus's survival outside the body include temperature, pH and moisture availability, whilst its possession of a lipid envelope (or not) will strongly influence its susceptibility to detergents and solvents like ethanol or isopropanol, which are commonly used as the basis for disinfectants. Viruses are not particularly heat resistant so pasteurization is a means by which they can be removed. Just as bacteria vary in their sensitivity to heat, so too do viruses, so it is difficult to quote a single temperature/time combination that is certain to destroy all viruses, nevertheless a temperature of 60 °C for periods between 1–10 hours has been found satisfactory for removing different viruses from liquid blood products, but removal of hepatitis from dried blood products requires temperatures of at least 80 °C and significantly longer exposures.

Viruses similarly vary in their sensitivity to extremes of pH. Most human pathogenic viruses are inactivated by acid, and exposure to pH 4 for 6 hours is sufficient to kill some sensitive viruses, whereas several days'

exposure may be required for more resistant species. Enteroviruses – those causing infection in the gastrointestinal tract – are transmitted by the faecal-oral route so they survive transient exposure to stomach acid at pH 1–3.

Ultraviolet (UV) light damages nucleic acids and so has the potential to kill all kinds of microorganisms. Viruses are at least as susceptible to UV light as bacteria, and both groups of organisms are readily killed by the germicidal lamps that are used to decontaminate both water (used for medicines' manufacture) and air (as in biological containment cabinets and operating theatres).

Enveloped viruses are generally much more sensitive to the effects of drying, detergents, organic solvents and disinfectants than nonenveloped ones because any process or chemical that damages or removes the lipid envelope is likely to render the virus particle noninfective, which, for practical purposes, is the same as dead. Despite their relatively greater sensitivity however, some enveloped viruses may survive on dry solid surfaces for a significant period of time; influenza virus, for example, despite its normal mode of transmission as an aerosol, has been shown to survive and remain an infection hazard for 24–48 hours on stainless steel or plastic, and hepatitis A (nonenveloped) can survive in dried faeces for 30 days or more. Table 6.1 shows the relative efficacies of some frequently used disinfectants against many of the common enveloped and nonenveloped viruses.

Acknowledgement

Chapter title image: PHIL ID #10145; Centers for Disease Control and Prevention.

Chapter 7

Characteristics of other microorganisms and infectious agents

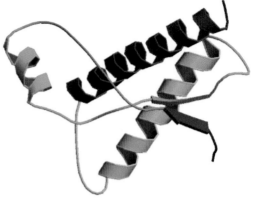

Prion protein, as found in bovine spongiform encephalopathy (BSE)

KEY FACTS

- Nonliving particles are capable of causing infections and include viroids and prions.
- Viroids are ss (single-stranded)RNA and prions are small proteins; only prions cause disease in humans.
- Chlamydias and rickettsias are both prokaryotic cells but differ from normal bacteria in that those species of importance are generally obligate intracellular parasites.
- Mycoplasmas are small bacteria devoid of cell walls. They are, however, capable of growing on cell-free medium.
- Mycoplasmas are resistant to cell-wall acting antibiotics such as the penicillins and cephalosporins.

Chapter 1 described the variety of microorganisms within the environment, while the major groups of bacteria, fungi, viruses and protozoa are dealt with in separate chapters. The purpose of this chapter is to briefly discuss those other clinically important microorganisms, which essentially lie outside these groupings and which might otherwise be overlooked.

Figure 7.1 gives rather a simplistic representation of the different agents under consideration showing the progression from nonliving chemicals through to complex, free living cell forms. It is not intended to suggest an evolutionary pathway.

7.1 Nonliving infectious particles

It is evident that chemicals and toxins can cause us harm but we do not tend to think of them as infectious agents. Being an infectious agent implies the ability to reproduce in the host and to be able then to transmit copies of that agent to another susceptible individual. Consequently, the term is often reserved for cellular structures which are capable of metabolism and are endowed with 'life'. However, when we considered viruses in Chapter 6 it

Essential Microbiology for Pharmacy and Pharmaceutical Science, First Edition. Geoffrey Hanlon and Norman Hodges.
© 2013 John Wiley & Sons, Ltd. Published 2013 by John Wiley & Sons, Ltd.

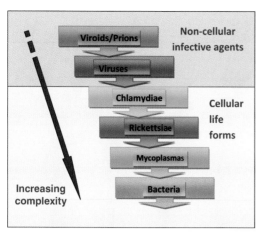

Figure 7.1   Schematic representation of the gradually increasing complexity from nonliving infectious particles through to free living microbes.

was apparent that our definition of living things isn't as straightforward as we first imagined.

However, viruses are not the simplest agents capable of causing infections and there are a number of simple chemicals which fall into this category.

7.1.1 *Viroids*

Viroids are small circular single-stranded RNA molecules which are not found complexed with protein. They are not infectious agents of humans and the most studied viroid has only 359 nucleotides (one tenth the size of the smallest known virus) and this causes disease in potatoes.

7.1.2 *Prions*

Prions are small self-replicating proteins which are devoid of nucleic acid. The smallest known prion contains only 250 amino acids. These agents are associated with a number of diseases:

- Creutzfeld–Jakob Disease (CJD) in humans.
- Scrapie in sheep.
- Bovine spongiform encephalitis (BSE) in cattle.

They are not infectious in the conventional sense. There are normal proteins in the membranes of the CNS which have the same amino acid composition as prion proteins but they have a different conformation. When prions are ingested they accumulate in the CNS membranes and by an autocatalytic process they convert normal proteins to abnormal ones. These abnormal

proteins cannot be destroyed by the body and lead to the destruction of CNS tissue in the brain.

Prions are highly resistant to heat and so are not destroyed by normal cooking or sterilization procedures. This is of concern because prions present in infected animals such as sheep and cows can enter the food chain and lead to disease in humans. In addition, any pharmaceutical products derived from animal products can pose a potential risk. A serious problem is the sterilization of surgical instruments which may have been used on a patient suffering from CJD. In this case the recommended treatments usually involve immersion in either 1N NaOH or 20 000 ppm hypochlorite for at least an hour and then autoclaving at 121 °C for one hour.

7.2 Viruses

Viruses have been dealt with in Chapter 6 and so no further information will be provided. The following characteristics are presented simply for completeness:

- Viruses are not cells.
- They are static structures – they have no metabolic ability of their own.
- They lack ribosomes and rely on host biosynthetic machinery for protein synthesis.
- Their genome is *either* DNA or RNA.

7.3 Chlamydiae

Chlamydiae are primitive small prokaryotic cells, which, when stained with the Gram staining procedure, appear Gram-negative. This is because their cell wall lacks the peptidoglycan present in most other bacteria. Their genome size is only about a quarter the size of that found in *E. coli*. They are obligate intracellular parasites which, unlike viruses, do possess some independent enzymes but lack an ATP generating system.

Unlike most bacteria, chlamydiae occur in two different morphological forms:

- A small elementary body (approximately 0.3 μm in diameter), which is the infectious form.
- A larger initial (reticulate) body (0.8 to 1.2 μm in diameter), which develops from the elementary body once inside the host cell. The initial body replicates by binary fission.

These agents are important pathogens for humans and cause a range of different disease states.

7.3.1 Trachoma

7.3.1.1 Eye infections

Chlamydia trachomatis is the most common *infective* cause of blindness in the world (the most common noninfective cause is cataract). It is primarily a problem in developing countries where lack of water and poor hygiene has led to approximately 500 million people being affected, of which 6 million are totally blind. The chlamydiae cause repeated infections of the eyelids and conjunctiva, which ultimately causes the eyelids to inflame and turn inwards. The eyelashes then continually abrade the surface of the eye resulting in scarring, further infection of the cornea and subsequent blindness. Despite their devastating consequences these infections are easily treated with antibiotics such as tetracycline, but such products are not readily available in many parts of the world where these infections are commonplace.

In developed countries the same organism can cause a milder form of conjunctivitis.

7.3.1.2 Genital tract infections

Different serogroups of the species *C. trachomatis* which causes trachoma are responsible for sexually transmitted infections of the urethra, cervix, fallopian tubes or uterus. These infections are quite widespread and it is estimated that in the US approximately 3 million people are affected each year.

7.3.1.3 Respiratory tract infections

C. psittaci can give rise to a form of pneumonia (psittacosis) usually acquired as a result of contact with infected birds. *C. pneumoniae* can cause a range of upper and lower RTIs such as sore throat, otitis media, bronchitis, pharyngitis, laryngitis and pneumonia.

7.4 Rickettsiae

These are small (0.7 to 2 μm), Gram-negative bacteria, which are pleomorphic – varying in shape from cocci to filaments depending on the environment (see red-stained cells in Figure 7.2). Most (but not all) are obligate intracellular parasites, which have a requirement for coenzyme A, NAD and ATP. Intracellular multiplication is by binary fission but is very slow and the generation time can be in the order of 8 hours. These agents are infectious pathogens and all except *Coxiella burnetii* are transmitted to humans via an arthropod vector (fleas, ticks and lice).

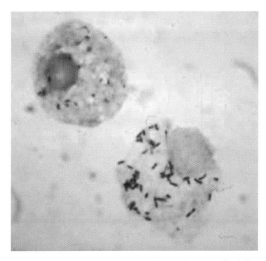

Figure 7.2   Gimenez stain of tick haemolymph cells infected with *Rickettsia rickettsii*. *Source:* Centers for Disease Control and Prevention (CDC); http://commons.wikimedia.org/wiki/File: Rickettsia_rickettsii.jpg.

Louse-borne infection
- Rickettsiae multiply in the cells of the gut wall of a louse and get into its faeces.
- Humans are infected by rubbing faeces or crushed louse into skin abrasions.
- Infection might also occur by inhalation of dried faeces.
- The louse does not bite!

Flea-borne infection
- Flea infected when it takes a blood meal from an infected animal.
- Rickettsiae pass into faeces.
- Humans are invaded by fleas and are infected through faeces in same way that they are infected by the louse.

Mite- and tick-borne infection
- An insect picks up a microorganism when it bites an infected host.
- It passes the microorganism onto humans when they are bitten subsequently.

7.4.1 Rickettsial diseases

Rickettsial disease is characterized by multiplication of the cells in the vascular endothelium (the inner lining of the blood vessels). They cause symptoms on the skin, CNS and liver and tend to persist in the body for extended periods of time. As with most arthropod infections there is no person

Table 7.1 The different rickettsial diseases and their characteristics.

Microorganism	Disease	Arthropod vector	Vertebrate reservoir	Severity of disease
R. rickettsii	Rocky Mountain spotted fever	Tick	Dogs, rodents	+
R. akari	Rickettsial pox	Mite	Mice	−
R. conorii	Mediterranean spotted fever	Tick	Dogs	+
R. prowazekii	Epidemic typhus	Louse	Human	++
R. typhi	Endemic typhus	Flea	Rodents	−
R. tsutsugamushi	Scrub typhus	Mite	Rodents	++
Coxiella burnetii	Q fever	None	Sheep, goats, cattle	+
Rochalimaea quintana	Trench fever	Louse	Human	+

to person spread of disease. A few important examples are given below and are summarized in Table 7.1.

7.4.1.1 Rocky Mountain spotted fever

- This chiefly occurs in rural populations whose work brings them into contact with wildlife and tick-infested stock.
- It may also be spread through ticks of domestic animals.
- A severe rash covers the entire body (Figure 7.3).
- This can develop into gangrenous necrotic lesions.
- Multiple organs may be affected.
- There can be up to 30% mortality.

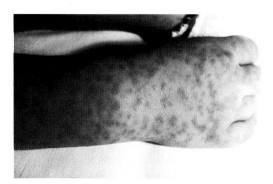

Figure 7.3 Child's hand showing typical rash of Rocky Mountain spotted fever. *Source:* PHIL ID #1962; Centers for Disease Control and Prevention.

7.4.1.2 Typhus (epidemic, murine and scrub)

- One of the classic epidemic diseases of history.
- Characterized by sudden onset, high fever, malaise, severe headache, generalized rash.
- It has 40% mortality.

7.4.1.3 Q fever

- Named after Q for query (coined in 1930s).
- Caused by *Coxiella burnettii*.
- Spread by deer ticks but can arise from sources other than insect bites, such as inhalation, ingestion and cuts.
- Acute febrile illness.
- Sudden onset, headache, pulmonary involvement.
- Complications – chronic endocarditis.

7.5 Mycoplasmas

These are the smallest prokaryotic cells capable of an independent existence and are therefore able to grow on standard microbiological media. However, they are quite fastidious in their nutritional requirements and do need complex growth media. The cells lack peptidoglycan in their walls; hence they have little rigidity and are variable in shape and size. Instead of the normal cell wall structure mycoplasmas have a triple layered membrane comprising proteins and lipids. Some species have cholesterol in their membranes, which is unusual as this is normally absent from bacteria and fungi and is only found in human membranes.

Table 7.2 Characteristics of different groups of infectious agents.

Characteristic	Bacteria	Mycoplasmas	Chlamydiae	Ricketssiae	Viruses	Prions	Viroids
Can lead independent existence	Yes	Yes	No	No	No	No	No
Independent protein synthesis	Yes	Yes	Yes	Yes	No	No	No
Ability to generate metabolic energy	Yes	Yes	No	No	No	No	No
Rigid envelope	Yes	No	Variable	Yes	Variable	No	No
Mode of reproduction	Binary fission	Binary fission	Binary fission	Binary fission	Variable and complex	None	None
Nucleic acid content	DNA and RNA	DNA and RNA	DNA and RNA	DNA and RNA	DNA OR RNA	None	RNA
Susceptibility to 'true' antibiotics[a]	Yes	Yes	Yes	Yes	No	No	No

[a]See Chapter 9.

There are three main genera of pharmaceutical interest, namely *Mycoplasma; Ureaplasma* and *Acholeplasma*. While a number of these are important pathogens of birds and animals *M. pneumoniae* is the most important human pathogen. It causes atypical pneumonia as well as genital-tract infections (nonspecific urethritis) and other joint and inflammatory conditions.

Because mycoplasmas do not contain any peptidoglycan in their cell walls the infections cannot be treated with antibiotics which act on this structure (principally, but not exclusively, the β-lactam antibiotics). However, mycoplasmas are susceptible to other antibiotics such as tetracyclines or erythromycin.

Table 7.2 summarizes the main features of the various infectious agents described in this chapter.

Acknowledgement

Chapter title image: Image by Lopez-Garcia F. et al., Research Collaboratory for Structural Bioinformatics (RCSB); http://commons.wikimedia.org/wiki/File: BOVINE_PRION_PROTEIN_1dx0_asym_r_500.jpg.

Part II

Microorganisms and the treatment of infections

Chapter 8
Infection and immunity

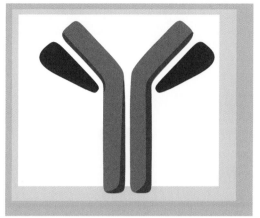

The pattern of protein chains found in antibody molecules

KEY FACTS

- Although we harbour huge numbers of microorganisms in and on our bodies we do not usually consider ourselves to be in any way infected.
- An infection occurs when an organism causes an adverse reaction in the host.
- Pathogenic microorganisms have acquired virulence factors to enable them to cause disease.
- Most pathogens cause disease via the production of toxins.
- The mode of transmission may be by a variety of routes including faecal, exhalation droplets, direct contact, animal bite and wound infections.
- The body's immune response comprises innate immunity and adaptive immunity.
- Innate immunity is multifactorial and includes phagocytic cells.
- The adaptive immune response occurs as a result of infection and comprises antibody production and cellular immunity.
- B and T cells resulting from initial infection are long lived and are responsible for immunological memory.
- Passive immunity involves inoculation with preformed antibodies.
- Active immunization requires vaccination with antigen to elicit a natural antibody response.

One of the main reasons we are interested in microorganisms is because they have the potential to harm us – to cause infections. However, we must keep this in perspective. *Bergey's Manual of Determinative Bacteriology* lists over 500 different bacterial genera but only about 10% of these contain species which may be capable of causing disease. Not every species within a particular genus can cause disease and neither can every strain within a species. Therefore the total microbial population able to cause infections in humans is actually very small.

8.1 What is an infection?

Table 8.1 shows examples of bacteria which are found as residents (normal microflora) in or on our bodies. These can be present in extremely high numbers, particularly in the bowel, but we would not describe ourselves as being infected; indeed the presence of these bacteria is of benefit to us. An infection is therefore defined as *an adverse reaction caused by the presence of a microorganism*. This

Essential Microbiology for Pharmacy and Pharmaceutical Science, First Edition. Geoffrey Hanlon and Norman Hodges.
© 2013 John Wiley & Sons, Ltd. Published 2013 by John Wiley & Sons, Ltd.

Table 8.1 Normal human microflora.

Oral cavity	Gastrointestinal tract
Staphylococci	*Escherichia coli*
Streptococcus mutans	Other Enterobacteriacae
Other streptococci	Enterococci
Spirochaetes	Yeasts
Actinomycetes	Actinomycetes
Bacteroides spp.	*Bacteroides* spp.
Fusobacteria	*Clostridium* spp.
Yeasts (*Candida*)	Bifidobacteria
	Eubacteria
Sinuses	**Genital tract**
Streptococcus pneumoniae	Lactobacilli
Other streptococci	Streptococci
Haemophilus influenzae	Corynebacteria
Actinomycetes	Mycoplasmas
Fusobacteria	Peptococci
Peptococci	Actinomycetes
Propionibacterium spp.	Yeasts
Throat	**Skin**
Staphylococci	Staphylococci
Streptococcus pneumoniae	Micrococci
Streptococcus pyogenes	Corynebacteria
Haemophilus influenzae	*Propionibacterium* spp.
Corynebacteria	
Fusobacteria	
Bacteroides spp.	
Candida spp.	

may occur as a result of damage caused by the microorganism or its products, or may arise from the host inflammatory response to the microorganism.

This leads us on to two further definitions which are important in our understanding of infections. A pathogen is defined as *a bacterium capable of causing an infection in a susceptible host*. For a number of bacteria this is not as straightforward as it might appear. For example, *E. coli* is found in large numbers in our bowel, but that does not constitute an infection. However, some strains of this bacterium are capable of causing profound disease (see Table 8.2). There are hundreds of different

serotypes of *E. coli* based upon their O (cell wall), H (flagella) and K (capsular) surface antigens hence leading to designations like *E. coli* O157:H7.

These strains have therefore acquired specific attributes not possessed by our gut flora which transforms them into harmful bacteria. These are usually attributes associated with the ability to adhere to epithelial cells (e.g. fimbriae) and the ability to produce toxins.

A different situation is illustrated by *Staphylococcus epidermidis*, which is a major component of the skin microflora. This is generally regarded as a nonpathogen, however, when the skin is breached by the introduction

Table 8.2   Examples of some pathogenic strains of *Escherichia coli*.

ETEC Enterotoxigenic *E. coli*
- Colonize the small intestine.
- Can produce 2 types of toxin; LT (labile toxin) and ST (stable toxin).
- Cause mild to severe traveller's diarrhoea.

EPEC Enteropathogenic *E. coli*
- Cause diarrhoeal disease in infants but do not produce LT or ST.
- Lack fimbriae but bind to host cells using intimin.
- Colonize the small intestine and can invade the mucosa.
- Can produce a cytotoxin but its role in disease is unclear.

EHEC Enterohaemorrhagic *E. coli*
- Colonize the colon using fimbriae and can invade mucosa.
- Produce a cytotoxin.
- Cause haemorrhagic colitis with watery, bloody diarrhoea but no fever.
- Haemolytic uraemic syndrome can lead to severe complications of kidney failure and death.

EIEC Enteroinvasive *E. coli*
- Invade and destroy the mucosal cells of the ileum and colon.
- Give rise to bacillary dysentery.

of implanted medical devices such as catheters, this organism can give rise to severe infections. In this case the bacterium has not acquired any specific attributes; it just finds itself in a different environmental situation and takes advantage of the opportunity.

The last definition is that for virulence, which is *the capacity of a pathogen to cause disease*. Virulence is a general term which reflects two main properties: infectivity and severity.

Infectivity can be quantified by determining the number of microorganisms required to cause an infection. As can be seen from Table 8.3, this varies from a single cell to tens of millions of cells. Again, we need to be a bit cautious how we interpret this information because in a cholera outbreak, for example, it would be very easy to consume hundreds of millions of vibrios present within a single drop of faeces as the patient will be excreting several litres of highly infected, liquid faeces each day. It must also be appreciated that infectivity is not necessarily an indication of severity – for example, *Mycobacterium tuberculosis* has a relatively low infectivity in that most people exposed to the organism do not go on to develop the disease. However, those that do contract tuberculosis suffer from a severe life-threatening infection. On the other hand the common cold virus has a very high infectivity but the disease which results is mild and self-limiting.

Needless to say, the most dangerous pathogens are those with high infectivity and high severity, such as smallpox.

Table 8.3   Estimates of the number of microorganisms required to cause an infection in a susceptible host.

Disease	Microorganism	Dose
Cholera	*Vibrio cholerae*	10^8 cells
Typhoid	*Salmonella typhi*	10^5–10^7 cells
Dysentery	*Shigella dysenteriae*	10^2 cells
Food poisoning	*E. coli* 0157	10–100 cells
Histoplasmosis	*Histoplasma capsulatum*	1 spore

Table 8.4 Examples of different virulence factors possessed by pathogenic bacteria.

Virulence factor	Examples	Mechanism of action
Pili/fimbriae	*Neisseria gonorrhoea* *Escherichia coli*	Aid attachment to epithelial cells
Capsules	*Klebsiella pneumoniae* *Streptococcus pneumoniae*	Aid attachment and resist phagocytosis
Exotoxins	Numerous – see Table 8.7	Cause various damaging effects to host cells
Endotoxins	Most Gram-negative pathogens	Release endogenous pyrogens causing fever, rash and haemorrhage, circulatory collapse and death
Leucocidin	*Staphylococcus aureus*	Kills phagocytic leucocytes
Coagulase	*Staphylococcus aureus*	Protects pathogen at site of infection by forming fibrin clot
Collagenase	*Clostridium perfringens*	Dissolves collagen of tissues allowing pathogen to spread from infection site
Lecithinase	*Clostridium perfringens*	Destroys host cell membranes
Hyaluronidase	*Streptococcus pyogenes*	Breaks down hyaluronic acid allowing pathogen to spread
Fibrinolysin	*Streptococcus pyogenes*	Dissolves fibrin clots formed by body defences and allows pathogen to spread

The basic metabolism of pathogenic and non-pathogenic bacteria is very similar but, clearly, those bacteria capable of causing disease possess additional attributes or virulence factors which allow them to cause infections. Examples of these virulence factors are shown in Table 8.4. Some permit the pathogen to attach to epithelial cells and establish an infection, others enable it to survive host defences and still others allow it to spread from the initial infection site.

Infections can take many different forms and Table 8.5 describes the different types of infection which can result from invasion by a virulent pathogen.

The pathogenicity of a bacterium involves the ability to:

- overcome the body's defence mechanisms;
- adhere to body surfaces and increase in numbers;
- produce toxic substances;
- move to other sites.

Whether or not a pathogen gives rise to an infection is a balance between the virulence of the organism and the efficiency of the host's nonspecific defence mechanisms. If an infection becomes established then the host's specific immune system is activated in an effort to eliminate the threat.

8.1.1 Overcoming the body's defence mechanisms

The body possesses a number of innate or nonspecific defence mechanisms in addition to the more specific immune system mechanisms involving antibodies and cellular immune responses. In order for a pathogen to initiate an infection these must first be overcome. The nonspecific defence mechanisms include:

- physical barriers (skin and mucous membranes);
- mechanical clearance mechanisms (mucociliary transport, peristalsis);
- chemical barriers (lysozyme, stomach acid);
- competition from resident microflora;
- phagocytosis.

More information on the nonspecific defence mechanisms of the body can be found in the text box.

Table 8.5 Examples of different types of infections.

Type of infection	Description	Examples
Primary	Single organism infecting an otherwise healthy host. Runs a characteristic course.	Cholera; pneumococcal pneumonia.
Secondary	Microbial invasion by a different organism following a primary infection. Variable course.	Bacterial pneumonia following viral lung infections.
Opportunistic	Infection caused by normal flora or transient bacteria when normal host defences are compromised.	*Staph. epidermidis* infections on implants. Burn wound infections e.g. *Acinetobacter baumanii*.
Acute	Rapid onset; brief duration.	Influenza.
Chronic	Prolonged duration.	Tuberculosis.
Localized	Confined to small area.	Staphylococcal boil.
Generalized	Spreads throughout the body.	Gram negative bacteraemia.
Pyogenic	Formation of pus.	Staphylococcal or streptococcal skin infections.
Fulminant	Infections that occur suddenly and overwhelm the patient.	Viral haemorrhagic fevers, e.g. Ebola.
Latent	Infecting agent remains dormant in body and infective episodes flare up intermittently.	Latent viruses such as Herpes simplex and Varicella zoster.

Additional information on nonspecific defences

Intact **skin and mucous membranes** provide a significant barrier to the ingress of microorganisms and, unless damaged in some way, most organisms will not be able to gain access to the body via this route. The respiratory tract and gastrointestinal tracts also have clearance mechanisms for those bacteria progressing into these areas. **The cough** is simply a reflex to expel any particulate irritants which find their way into the throat. As it is a protective mechanism we should be cautious about suppressing the cough using drugs. **Mucociliary transport** comprises ciliated cells in the lining of the respiratory system; these beat in unison and transport a carpet of mucus from the lower airways up into the throat to be swallowed. Inhaled microorganisms become trapped on the sticky mucus and end up in the stomach, where they are destroyed. Smoking damages the ciliated epithelia leaving the patient more susceptible to lung infections. Increasing the rate of **peristalsis** is a mechanism for rapidly removing harmful gut pathogens. A number of body secretions including saliva contain the enzyme **lysozyme**, which destroys the peptidoglycan in bacterial cell walls. The **acid** of the stomach is also a powerful barrier to the passage of swallowed bacteria into the GI tract. The microorganisms (shown in Table 8.1) which comprise the **normal microflora** of the body act as a major impediment to colonization by invading pathogens. These are already well established in their particular niches and often produce inhibitory substances such as organic acids. **Cellular clearance mechanisms** include macrophages whose role is to engulf any invading foreign particles.

8.1.2 Adherence to body surfaces and increase in numbers

Having overcome any innate defence mechanisms, attachment of the pathogen to body surfaces is the first step in the infection process. If it cannot attach, it cannot establish itself and increase in numbers to a critical point where it starts to elicit an adverse effect. There are a number of different mechanisms by which specific pathogens attach to surfaces and initiate infections:

- Attachment and multiplication only on the surface of the mucosal epithelial cells (for example, *Vibrio cholerae*).
- Attachment to mucosal surface and then penetration and multiplication in the epithelial cells (for example, *Shigella dysenteriae*).
- Passage through the epithelium and spread into the deeper tissues *via* the circulatory system (for example, *Streptococcus pneumoniae*).

8.1.3 Production of toxic substances

It is reasonable to wonder how bacteria, which are so small, can cause us such harm and perhaps even kill us. In most cases it is not the cells themselves that cause the damage but the products or toxins that they produce. Broadly speaking these can be divided into two groups – endotoxins and exotoxins – the characteristics of which are shown in Table 8.6.

8.1.3.1   Endotoxins

Gram negative (but not Gram positive) cells possess lip-opolysaccharides (LPS) in their cell walls and this highly toxic material can be shed into the environment. This can occur while the cells are alive, but more importantly also when they die. LPS (also called endotoxins or pyrogens) are very heat stable and cause a range of toxic effects. The toxic effects of endotoxins are described in Chapter 3, which can be referred to for further information.

8.1.3.2   Exotoxins

Table 8.7 gives a few examples of some of the protein exotoxins produced by pathogenic bacteria. As can be seen, they are a very diverse group of molecules having an extensive range of effects, and they represent some of the most powerful poisons known.

8.1.4 Movement to other sites

As can be seen above, some pathogens, such as *Vibrio cholerae,* remain attached to the primary infection site and produce their adverse effects simply by producing toxins. Others such as *Shigella dysenteriae* give rise to limited penetration into the epithelial cells lining the gastro-intestinal tract and it is this that causes significant damage. There are, however, a number of pathogens which enter the body by one route and then move by various means to other sites, perhaps even the blood stream. Table 8.4 gives some examples of the enzymes that are produced by certain bacteria to enable them to do this. Often these enzymes are highly destructive to body tissues.

8.2 Mode of transmission of disease

In order to cause an infection a pathogen must travel from its usual reservoir (where it normally resides in the

Table 8.6   Comparison of the characteristics of endotoxins and exotoxins.

Endotoxins	Exotoxins
Gram-negative cells only	Gram-positive and Gram-negative cells
Found in pathogens and nonpathogens	Produced by pathogens only
Released when cells die	Secreted by living cells
Lipopolysaccharide component of cell wall	Protein
Heat stable	Heat labile
Limited activity mediated by release of cytokines	Very variable in their toxicity

Table 8.7 Examples of different bacterial exotoxins.

Disease and causative agent	Nature of toxin	Effect of exotoxin
Botulism *Clostridium botulinum*	Neurotoxin acts on motor neurones blocking acetylcholine release thus preventing muscle excitation.	Blocks nerve impulses in a state of relaxation.
Cholera *Vibrio cholerae*	Enterotoxin stimulates adenylate cyclase activity leading to reduced adsorption of Na^+ and Cl^- and increased secretion of bicarbonate.	Secretion of large amounts of water into the colon.
Diphtheria *Corynebacterium diphtheriae*	Sub unit cytotoxin made up of two parts. Fragment B facilitates entry of fragment A into cell.	Interferes with protein synthesis. Causes damage to heart, nerves and liver.
Dysentery *Shigella dysenteriae*	Subunit shiga toxin. Has enterotoxic, neurotoxic and cytotoxic activity.	Binds to ribosomes and inhibits protein synthesis. Causes neurological impairment.
Food poisoning *Staphylococcus aureus*	Food contains preformed heat stable enterotoxin.	Stimulates vomiting centre in the CNS.
Gas gangrene *Clostridium perfringens*	Multiple toxins produced including phospholipase C (α-toxin).	Causes necrosis of affected tissue.
Gastroenteritis *Escherichia coli*	Produce shiga-like toxin known as verotoxin; also enterotoxins.	Secretion of large amounts of water into the colon.
Pertussis (whooping cough) *Bordetella pertussis*	Tracheal cytotoxin paralyses cilia, and pertussis subunit toxin interferes with cAMP-regulated events.	Causes necrosis of the epithelial lining of the upper respiratory tract.
Pseudomembranous colitis *Clostridium difficile*	Enterotoxin (toxin A) and cytotoxin (toxin B). Glucosyltransferases that inhibit GTPases.	Cause mucosal damage and diarrhoea.
Scarlet fever *Streptococcus pyogenes*	Three similar pyrogenic and erythrogenic toxins.	Toxins injure capillaries and cause rash. Also stimulate macrophages producing cytokines.
Tetanus *Clostridium tetani*	Neurotoxin acts on spinal cord causing continual excitation of motor neurones.	Nerves are paralysed in a state of contraction.
Toxic shock syndrome *Staphylococcus aureus*	Superantigen stimulates the release of large amount of interleukins and tissue necrosis factor.	Causes rash, fever and shock.

environment) to a susceptible host. For many pathogens the normal reservoir is another human and the bacteria may escape in faeces, salivary droplets, skin exudates, blood and so forth. Once released from the reservoir, the pathogen may be transmitted by direct or indirect routes. An important factor in transmission is how long the pathogen can survive in the environment outside the host.

8.2.1 Faecal contamination

The intestinal tract harbours billions of bacteria, most of which are harmless. However, a number of important pathogens can infect the bowel and the faeces then act as a major source of infection, particularly if the patient suffers from profound diarrhoea. The following are some examples of pathogens transmitted via faecal spread.

Typhoid fever: Salmonella typhi
- Bacteria are shed in faeces of asymptomatic carriers and faeces/urine of patients with active disease.
- Some untreated patients shed organisms for many months. Organisms are localized in the gall bladder and these are chronic typhoid carriers.
- Inadequate hygiene spreads the organism to communal food and water supplies.
- Flies spread disease from faeces to food.

Other Salmonella infections: Salmonella typhimurium
- Epidemiology of disease is more complex than for *S. typhi*.
- Many infections are acquired by direct and indirect contact with infected animals and food.
- Salmonellae are found in poultry, eggs and raw milk.

Cholera: Vibrio cholerae
- Spread by ingestion of water, seafood and other contaminated foods.
- Organisms are shed from symptomatic and asymptomatic patients.
- Proper sewage disposal and maintenance of clean water supplies are essential in controlling cholera.

Bacillary dysentery: Shigella dysenteriae
- The main source of infection is excreta of infected and convalescent patients; true long-term carriers are rare.
- Direct spread is by the faecal/oral route, indirect spread is by contaminated food and inanimate objects. Flies serve as mechanical vectors.
- Epidemics occur in overcrowded areas with poor sanitation.

8.2.2 Exhalation droplets

Each time we cough, sneeze or even just talk, clouds of minute salivary droplets are expelled from our mouths and these will contain the bacteria resident in the respiratory tract at that time. Larger droplets will probably fall to the ground and contaminate surfaces close by. The moisture in smaller particles will evaporate very quickly and the residue which comprises proteins and viable bacteria are known as droplet nuclei. Being very small they will remain suspended for significant periods of time, can travel on air currents and may even be small enough to be inhaled. Particles greater than 5 μm will, if inhaled, impinge on the mucus layer which lines the upper respiratory tract and be expelled via the mucociliary transport system. Smaller particles (<5 μm) will remain airborne in the respiratory tract and may find their way into the lower reaches of the lung where they could cause disease.

The following examples are amongst the more important of those pathogens that are transmitted via exhalation droplets:

Streptococcal infections: Streptococcus spp
- 40–70% of the population carry streptococci in their throats.
- *Strept. pneumoniae* reaches the lung through inhalation.
- It lodges in alveoli and sets up an inflammatory reaction.

Diphtheria: Corynebacterium diphtheriae
- Humans are the only reservoir of infection.
- The disease is spread by contact with infected patients and carriers.
- More problematic in crowded institutions. Improvement in social conditions/vaccination has reduced incidence.
- Most UK cases now come from overseas.

Tuberculosis: Mycobacterium tuberculosis
- The most common form is pulmonary infection, which is highly infectious and is acquired by inhalation.
- Bacteria multiply within lesions in the lung called tubercles. These discharge into bronchi spreading disease to other parts of the lung and the environment.
- Indirect spread via inanimate objects is rare.

Meningococcal meningitis: Neisseria meningitidis
- Meningococcus found in the nasopharynx of 25% of the population.
- It is spread by respiratory droplets and close contact.
- It is not known why a small percentage of carriers go on to develop disease.

8.2.3 Direct contact

A small number of pathogens have their route of entry into the body via the skin or mucous membranes. These include a variety of occupational diseases and the sexually transmitted diseases. Some of these organisms cannot survive for long periods of time in the environment, hence the need for direct contact.

Anthrax: Bacillus anthracis
- An occupational disease of farmers, vets and people who handle hides and skin.
- The primary reservoirs are goats, sheep and cattle.
- Cutaneous anthrax – 95% of cases, low mortality.
- Pulmonary anthrax – 5% of cases, invariably fatal.

Brucellosis: Brucella abortus
- The reservoir is cattle – organisms are shed in milk.
- It is acquired by direct contact with tissues or ingestion of milk.
- It is an occupational disease of agricultural workers.

Syphilis: Treponema pallidum
- Transmitted by sexual contact – bodily contact is sufficient.
- The organism can enter the body via mucous membranes or the skin.
- Most infectious patients are those with untreated lesions.
- A mother can pass the disease to a developing foetus.

Gonorrhoea: Neisseria gonorrhoea
- The disease is spread by sexual contact.
- A mother can pass the disease to a developing foetus.
- Some people may be symptomless carriers for weeks or months.

8.2.4 Animal bite

This group comprises those diseases which are transmitted via animal vectors (mainly insects), and could include important infections such as malaria, rabies, sleeping sickness, yellow fever and dengue fever. The main bacterial and rickettsial diseases spread by insect vectors are given below. Normally, the insect will draw a blood meal from an infected host (often an animal) and then pass the infection on when taking a subsequent blood meal from a human.

Rocky mountain spotted fever, Q fever and typhus
These rickettsial diseases were discussed in Chapter 7 which can be referred to for more information.

Relapsing fever: Borrelia recurrentis
- Widespread in Africa and the Middle East.
- Epidemics occur when normal hygiene breaks down.
- The last epidemic in Europe was during World War II: 50 000 deaths.
- The epidemic form transmitted by body lice.
- The endemic form is transmitted by tick bites.

Lyme disease: Borrelia burgdorferi (spirochete)
- An arthritic illness first reported in Old Lyme, Connecticut, 1975.
- It is the most common tickborne infection on the Northern hemisphere.
- US 4000 cases per annum, UK 300–400 cases per annum.

8.2.5 Wound infections

A wide range of microorganisms can lead to infection of wounds depending on the environment. These will include the clostridia, which can infect dirty wounds such as might occur in war zones, or pseudomonads or staphylococci, which may infect wounds in hospitals. These are too diverse to be dealt with exhaustively here.

Leptospirosis (Weil's disease): Leptospira icterohaemorrhagia
- Reservoirs are pigs, dogs and rodents. Spirochetes are excreted in the urine of infected animals.
- It infects humans through minor cuts and scratches.
- It is an occupational disease among workers in frequent contact with contaminated water, such as in sewers, canals or fish markets. It also arises through recreational swimming in lakes and rivers.
- From an initial wound entry the organism disseminates to give infectious jaundice.
- The death rate is 2–10%. Death is usually due to liver or kidney failure and myocarditis.

8.3 Immune response to infection

8.3.1 Cellular components of the immune system

We have previously indicated that the human host is protected from infection by both a nonspecifc or innate immunity and also a specific or adaptive immune response. Innate immunity does not require an infection and comprises, among other things, phagocytic cells called macrophages which are able to recognize and engulf a wide range of microorganisms entering the body. Adaptive immune responses are triggered by infection and can bring about lifetime immunity to reinfection by that pathogen. This latter effect is mediated by lymphocytes and centres on the production of antibodies in response to the presence of antigens. The adaptive immune response is also pivotal in the protection provided by vaccination. Figure 8.1 shows the origins of those cells involved in the immune system. Whole textbooks have been written on the subject of immunology and so it is impossible here to give more than a highly truncated summary of a few important points.

Bacterial cells have molecules on their surface, which can bind to receptors on the surfaces of macrophages and neutrophils. This triggers these phagocytic cells to engulf the bacterium and also to release chemical mediators such as cytokines. These cytokines modify the behaviour of other cells and can also bring about inflammation. Inflammation is characterized by pain, redness, swelling and a local increase on temperature, all of which are due to the action of cytokines on local blood vessels. The phagocytic cells which form the innate immune system play a crucial role in defending the body against infection by microbial pathogens but some pathogens have evolved mechanisms to avoid them. In addition, viruses do not possess the surface molecules which phagocytic cells can recognize and so they may also evade engulfment. Consequently, if the phagocytic cells cannot eliminate the pathogen, an infection will result.

8.3.2 Clonal selection and immunological memory

At this point the adaptive immune response in the form of lymphocytes comes into play. Each individual lymphocyte has the capacity to recognize a single antigen, which might at first sight seem rather limiting. However, there are millions of circulating lymphocytes and each one has a different recognition capability. During development in the bone marrow, the progenitor cell gives rise to a large number of lymphocytes each with a different recognition capability. Those lymphocytes which recognize self-antigens are at this stage eliminated. What remains are those cells which respond only to foreign antigens. When the lymphocyte encounters an antigen specific for the receptor it carries, the cell proliferates to produce large numbers of identical cells, termed clones, which differentiate into cells capable of producing antibody specifically directed against the antigen which elicited the response. This is known as clonal selection and the process of clonal expansion takes about five days to complete.

Some of the cells which form the clone remain in the system after the antigen has been eliminated. This is what

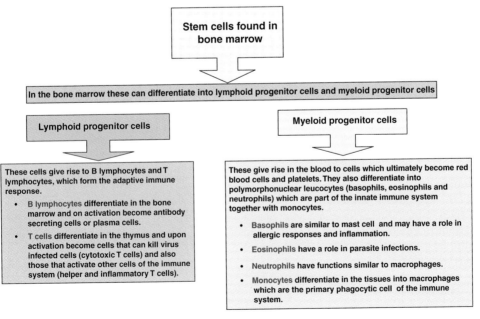

Figure 8.1   The origins of the cells involved in the immune system.

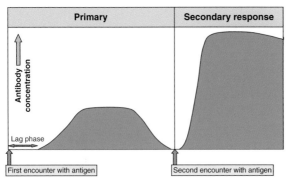

Figure 8.2   Antibody production following encounters with an antigen.

gives rise to immunological memory – the capability to give a more rapid and enhanced response when the body encounters the antigen for a second time (see Figure 8.2).

8.3.3 Antibody structure and function

Antibodies are also known as immunoglobulins and are made up of 4 polypeptide chains; two identical small chains (called light chains) and two identical large chains (called heavy chains). These are assembled together into a Y shaped structure linked by disulphide bridges as shown in Figure 8.3.

Each end of the Y-shaped molecule contains a variable region which is a receptor site for a specific antigen. The antibody can eliminate antigens (invading microorganisms or their toxic products) by binding to them and thus preventing them from acting on host cells. This is called neutralization. Sometimes this is not effective and so a second mechanism is opsonization. Here the antibody coats the antigen enabling the phagocytic cells to recognize the constant region of the antibody molecule and thus destroy the antigen. The constant region is at the

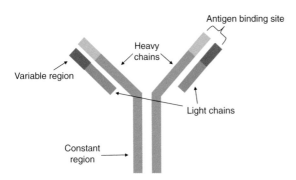

Figure 8.3   Antibody structure.

bottom of the two heavy chains. The final function of antibodies is to activate the complement cascade. The complement cascade is a series of plasma proteins which are sequentially activated resulting in components which can kill bacteria by lysis. There are different forms of immunoglobulins (Ig for short) and these may take the form of individual molecules, shown in Figure 8.3, or they may form dimers or pentamers (five molecules in a cluster). Some of the characteristics of these immunoglobulins are given below:

- IgA – dimer. Located in mucosal areas of the body such as the gut, respiratory tract and urogenital tract. Also found in some body fluids such as tears, saliva and breast milk. Prevent colonization of mucosa by pathogens.
- IgD – monomer. Activates basophils and mast cells to produce antimicrobial agents to destroy pathogens.
- IgE – monomer. Involved in allergic response. Binds to allergens and causes release of histamine from mast cells and basophils. Also provides protection against worm infestation.
- IgG – monomer. Provides the bulk of the antibody protection against invading pathogens. It is the only immunoglobulin capable of crossing the placenta and so conferring protection to the foetus.
- IgM – pentamer. Provides protection against invading pathogens while the concentrations of IgG are still low in the early stages of infection.

8.4 Vaccination and vaccines

We have already seen that when a pathogen evades the innate immune response an infection can result and at that point the adaptive immune response kicks in. However, this process is relatively slow and the initial antibody titres are quite low. For highly virulent pathogens the delay in response may prove to be extremely dangerous for the patient. On a second encounter with that same pathogen the adaptive immune system reacts much faster and to a higher level. This is due to immunological memory and people who have survived an infection with a pathogen are often immune to any subsequent reinfection. There are exceptions to this including highly antigenically unstable viruses like influenza, which mutates frequently; patients who are immunocompromised; and latent infections such as herpes simplex virus. In addition it is not unusual for a person to suffer from multiple episodes of the common cold each year. This is

Table 8.8 Examples of types of vaccines.

Vaccine Type	Viral infections	Bacterial infections
Live (attenuated)	Yellow fever Measles Mumps Rubella Polio	BCG (to prevent tuberculosis) Typhoid
Killed (inactivated)	Influenza Polio	Cholera Pertussis (whooping cough) Typhus
Toxoid	None	Diphtheria Tetanus

not a failure of the immune system – it is simply that there are over 100 antigenically distinct strains of the rhinovirus responsible and immunity to one strain does not protect us from the others.

From this it follows that it should be possible to artificially induce an 'infection' in a patient as a prophylactic measure in order to ensure that when they meet the pathogen for real the immune system will recognize it immediately and respond effectively. This is the basis of vaccination although, of course, we need to make sure that the patient isn't exposed to the live, virulent pathogen first time round. The first vaccination (and the origin of the term) was carried out by Edward Jenner in the late eighteenth century.

Table 8.9 Characteristics of vaccine types.

Vaccine type	Live (attenuated)	Killed (inactivated)
Route of administration	May be oral.	Usually injection.
Doses	Usually single dose.	Usually multiple doses.
Adjuvant (included in formulation to enhance recipient's immune response)	Not required.	Usually necessary.
Duration of immunity	Years – throughout life.	Months to years.
Immune response	IgG; IgA; IgM; cell mediated.	IgG; little or no cell mediated.
Advantages	Mimics natural infection, so more effective. Exposure through natural route – more appropriate (perhaps localized response).	No chance of active infection.
Disadvantages	Causes active (mild) disease that may be transmitted to others. Response may be unexpectedly great if recipient has poor immune function.	Usually less effective than live vaccines. Require repeat administration.

He immunized a child with cowpox (vaccinia) virus as protection against the closely related but highly dangerous smallpox (variola) virus. This is termed active immunization, as exposure to the antigen causes the body to produce the required B and T cells. Passive immunization is the process whereby artificially prepared antibodies are injected into a patient to provide immediate protection from a particular threat. Here the response is rapid but short lived as there are no B and T cells to continue production. Passive immunization is used in emergency situations where the patient may have been exposed to a toxin (snake bite) or a virus such as rabies.

Vaccines can be of a number of different types:

- Live (attenuated) – consists of live pathogen, but its virulence has been markedly reduced.
- Killed (inactivated).

- Component
  - bacterial cell components – for example, surface polysaccharides;
  - viral subunits – for example, capsid proteins;
  - peptide vaccines – recombinant peptides and proteins;
  - DNA vaccines – plasmid DNA encoding relevant antigen gene.
- Toxoid – toxins which have been modified to remove toxicity without affecting immunogenicity.

Examples of some of the bacterial and viral infections for which the different types of vaccine outlined above are used are shown in Table 8.8.

The characteristics of live (attenuated) and killed (inactivated) vaccines are quite different and each has their advantages and disadvantages. The main characteristics of live and killed vaccines are given in Table 8.9.

Chapter 9

The selection and use of antibiotics

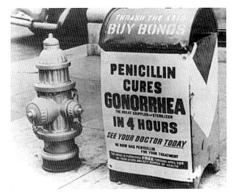

1944 US poster

KEY FACTS

- An antibiotic must exhibit selective toxicity: it should kill, or inhibit the growth of, a microorganism, without harming human cells.
- It usually does this either by attacking unique targets that only exist in microbial cells or by having a higher affinity for the bacterial version of the target than the human one.
- The major antibiotic targets are bacterial ribosomes and enzymes that make folic acid, DNA or peptidoglycan in the bacterial cell wall.
- The term 'spectrum of activity' describes the range of organisms against which the antibiotic acts. Very few antibiotics exhibit activity against more than one out of the three groups of organisms, bacteria, fungi and protozoa.
- All antibiotics exhibit some degree of toxicity; this may be sufficiently severe to require routine blood-level monitoring of the drug in question.
- Injection-only antibiotics are largely restricted to hospital use.
- Antibiotics for immunocompromised patients should be bactericidal rather than merely bacteriostatic.
- Antibiotics should also have good penetration into tissues and macrophages, minimal interaction with other drugs and minimal susceptibility to resistance development.
- There are a few situations in which antibiotics are used in combination, but generally the preference is to use a single antibiotic with a narrow spectrum of activity rather than broad-spectrum drugs.
- The use of antibiotics for prophylaxis is not generally recommended except for minimizing infection during surgery.

In order to use antibiotics effectively, it is necessary not just to know the characteristics of the drugs themselves but to understand the factors that determine which antibiotic is best suited for a particular infection. The spectrum of activity (the range of organisms against which the antibiotic is effective), its toxicity, dosage frequency, susceptibility to resistance development and several other factors will influence the choice of antibiotic. Some of these will, in turn, be determined largely by the mode of action of the drug itself – the mechanism by which it kills, or inhibits the growth of, the infecting organism. This chapter therefore, will focus on these

Essential Microbiology for Pharmacy and Pharmaceutical Science, First Edition. Geoffrey Hanlon and Norman Hodges.
© 2013 John Wiley & Sons, Ltd. Published 2013 by John Wiley & Sons, Ltd.

features of antibiotic therapy and, in addition, consider two aspects of the use of antibiotics that have changed significantly in recent years: the use of antibiotic combinations and of antibiotics intended for prophylaxis rather than treatment of infections.

9.1 Mechanisms of antibiotic action

The most fundamental characteristic that an antibiotic must possess is the ability to kill or inhibit bacterial growth without harming the patient; in other words, it should have selective toxicity. It must, therefore, exploit the differences that exist between bacterial and human cells in order to kill one and leave the other unharmed; this can be achieved in different ways, the two most common of which are for the antibiotic to:

- interfere with the synthesis or function of a vital chemical or structure which exists in the bacterial, but not the human, cell (Figure 9.1);
- have a higher affinity for the bacterial version of an enzyme or cellular structure (for example, ribosome) than for the human form (because slight structural differences often exist between corresponding enzymes from different species).

Other strategies to achieve selective toxicity do arise, but they are relatively rare compared with the two above. There are, for example, differences in permeability that permit some antibiotics to enter bacterial cells far more readily than human ones. This is seen in the aminoglycosides, which are very water-soluble, cationic drugs and so

have difficulty crossing the phospholipid membrane of human cells by passive diffusion, but which are actively transported into bacteria. Tetracyclines, too, are actively accumulated by bacteria but not by human cells.

9.1.1 Antibiotics interfering with the bacterial cell wall and membrane

The cell wall is a very clear example of a structure that exists in bacteria but not in humans. The peptidoglycan component of the wall is essential to protect bacteria from osmotic lysis (Chapter 3). If it is not synthesized properly, water may enter the cell by osmosis and the membrane stretches until, like an overinflated balloon, the cell bursts and dies. Thus, any chemical that can interfere either with the synthesis of peptidoglycan, or with its structure or function after it has been synthesized, is likely to be much more toxic to bacteria than humans. All the β-lactam antibiotics (penicillins, cephalosporins, carbapenems and others) together with the glycopeptides (vancomycin and teicoplanin) and some less important antibiotics like bacitracin and cycloserine, interfere with aspects of peptidoglycan synthesis, and the enzyme lysozyme (found in tears and saliva) will degrade the polymer once it is formed. It follows from this that bacterial pathogens which do not possess peptidoglycan in the wall (for example *Mycoplasma* species) are intrinsically resistant to all these agents.

The cell membrane is a much more difficult structure to target by antibiotics because it is chemically similar in all types of cells so there are few differences to exploit. As a consequence, there is only a single group of antibacterial antibiotics, the polymyxins, which act by interfering

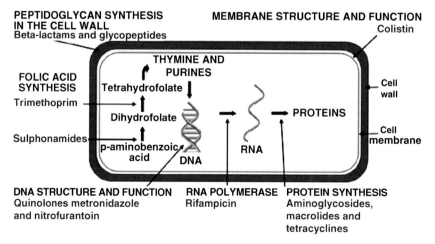

Figure 9.1 Sites of antibiotic action in the bacterial cell.

with membrane function (although there are several antifungal agents that target the membrane – Chapter 11). Colistin is the only systemically-used antibiotic in the polymyxin group, though others, like polymyxin B, are used topically. These antibiotics interact with phosphatidylethanolamine, a lipid present in much higher concentrations in Gram-negative cell membranes than in those of mammals or Gram-positive bacteria, to destabilize the membrane and cause fatal leakage of vital intracellular components.

9.1.2 Antibiotics interfering with folic acid synthesis and metabolism

Folic acid is a B-group vitamin, which, in its reduced form, is required to transfer methylene groups between biologically important molecules. A lack of folic acid leads to reduced synthesis of (amongst other things) thymine and purine bases; this, in turn, results in diminished nucleic acid production and ultimately cell division is slowed or stopped. Folic acid is essential for normal metabolism in all types of cell, but the means by which it is obtained differs: most bacteria synthesize it from its three component molecules (pteridine – an organic base, para-aminobenzoic acid (PABA) and glutamic acid) and they cannot obtain it from the extracellular fluid. Humans, on the other hand, cannot synthesize it and must obtain it from the diet. This difference is crucial because it is the basis for the selective toxicity of the sulphonamides. This large group of antimicrobial drugs have a common structural feature in that they all resemble the PABA (Figure 9.2) that bacteria use to synthesize folic acid, and so they inhibit the enzyme concerned. Because humans require the preformed vitamin they do not possess a corresponding enzyme, so their folic acid availability is unaffected.

Folic acid itself is inactive, and it needs to be reduced first to dihydrofolate and then to tetrahydrofolate, which is the biologically active form; these reactions occur in all types of cell. Several drugs interfere with the action of dihydrofolate reductase, notably the antibacterial drug trimethoprim, pyrimethamine (an antimalarial), and methotrexate, which is used in cancer chemotherapy. Trimethoprim and pyrimethamine owe their selective toxicity to the fact that they have a very much greater affinity for the versions of dihydrofolate reductase found in bacteria and the malarial parasite respectively than for the human version of the enzyme.

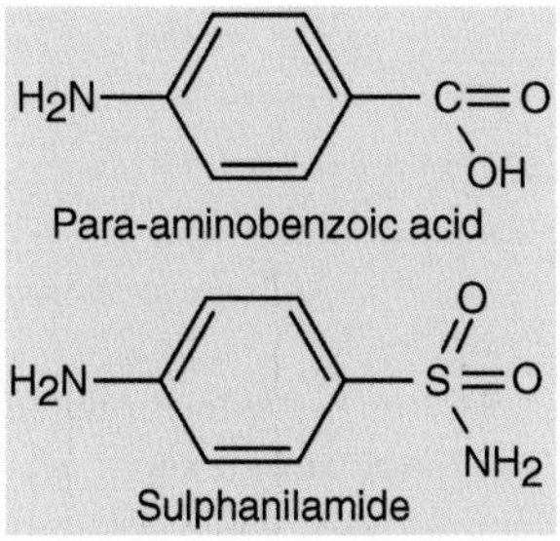

9.1.3 Antibiotics interfering with nucleic acids

In contrast to humans which possess 46 linear chromosomes, bacteria normally have a single circular chromosome, which is approximately 500–1000 times the length of the bacterial cell itself; it therefore needs to be tightly coiled to fit into the available space. Enzymes that alter the shape and unwind the helical structure of DNA occur in all cells and are called topoisomerases, but the supercoiling of DNA that is required by bacteria is achieved by the action of a specific topoisomerase which is termed DNA gyrase, and it is this enzyme which is the target for the fluoroquinolone antibiotics; again, it is unique to bacteria.

The basic structure of DNA is, of course, similar in all types of cell, so antibiotics that specifically damage bacterial DNA without also damaging the DNA in the cells of the human patient are not common. However, there are antibiotic pro-drugs, notably metronidazole and nitrofurantoin, which, themselves, are inactive, but become activated after reduction inside microbial cells to produce free radicals and metabolites that cause DNA strand breakage and destabilization of the DNA helix respectively. Metronidazole is reduced in any cell with a sufficiently low redox potential, so it is effective against any obligate anaerobe – not just bacteria but protozoa too. Mammalian cells and bacteria growing aerobically do not create sufficiently reducing conditions for the activation of metronidazole to occur and so they remain unharmed. Nitrofurantoin is reduced by bacterial flavoproteins so it is effective against anaerobes and aerobic bacteria such as those causing urinary tract infections for which the drug is most commonly employed.

The rifamycins (rifampicin and rifabutin) also interfere with nucleic acid synthesis, but in this case it is RNA production that is affected because the drugs specifically bind to bacterial DNA-dependent RNA polymerase (the enzyme that transcribes the genetic code from DNA to RNA).

Para-aminobenzoic acid

Sulphanilamide

Figure 9.2 Structural similarities between para-aminobenzoic acid and sulphanilamide (the first sulphonamide).

9.1.4 Antibiotics interfering with protein synthesis

The process of protein synthesis and the ribosomes themselves are the most common targets for antibiotics, largely because bacterial and eukaryotic ribosomes differ significantly in size and composition: 30s and 50s ribosomes occur in bacteria, whereas human cells have the larger 40s and 60s ribosomes which contain different structural proteins and, therefore, different antibiotic binding properties (Figure 3.8). The list of drugs interfering with ribosome function and protein synthesis includes the important, relatively large groups like the tetracyclines, aminoglycosides and macrolides, as well as individual antibiotics having limited or specialist applications, for example chloramphenicol, clindamycin, fusidic acid, mupirocin, linezolid and Synercid™. Again, it will not be possible to describe the action of each of these in detail here and only the major groups will considered.

- The aminoglycosides (gentamicin, amikacin, neomycin and others) have at least two effects on protein synthesis, which arise at different concentrations: at subinhibitory (nonlethal) levels they cause misreading of the genetic code so that incorrect amino acids are inserted into proteins, which, as a result, are nonfunctional. At higher concentrations they irreversibly bind to the 30S ribosome and prevent initiation of protein synthesis and elongation of peptide chains (and have other effects too, for example, membrane disruption).
- The tetracyclines reversibly bind to the 30S ribosome and inhibit binding of transfer RNA to the acceptor site on the 70S ribosome.
- The macrolides and structurally similar azalides (erythromycin, azithromycin, clarithromycin and telithromycin) inhibit translocation of the peptidyl tRNA from the A to the P site on the ribosome by binding to the 50S subunit.

9.2 Factors to consider in selecting an antibiotic

It is useful here to identify the desirable characteristics of an antibiotic. Ideally it should:

- possess activity against the widest possible range of organisms;
- exhibit minimal toxicity for the human (or animal) patient;
- not destroy 'friendly bacteria';

- have good oral absorption;
- have favourable pharmacokinetics, particularly a long half life, and be unaffected by pH and oxygen availability;
- exhibit bactericidal rather than merely bacteriostatic activity;
- exhibit good penetration into tissues and cells (including macrophages);
- not interact with other drugs;
- not be prone to resistance development;
- be inexpensive.

Some of the above are so self-evident that they need little or no further explanation, but more detail is justified in some cases.

9.2.1 Spectrum of activity

An antibiotic is usually effective against just one category of microorganism; rarely does an antibiotic work against both fungi and bacteria for example, and even amongst bacteria, many antibiotics preferentially show activity against just one group, for example Gram-positive or Gram-negative species, mycobacteria or anaerobes. Broad-spectrum antibiotics, such as amoxicillin, are often prescribed both for infections where the causative organism is not routinely identified, for instance urinary tract infections, and in more serious cases when the identity is still to be confirmed by the laboratory. They offer the obvious advantage that they have a higher probability of being effective than a narrow-spectrum agent, but this versatility has led to more widespread use and an associated increase in resistance. For example, E. coli resistance to ampicillin as a result of a specific β-lactamase enzyme was first described in 1965, but by the year 2000 approximately half of both hospital- and community-isolated strains of E. coli possessed the gene in question and were resistant to both ampicillin and amoxicillin as a consequence.

9.2.2 Toxicity and side effects

Antibiotics vary in their toxicity and side-effect profiles and some, like the aminoglycosides and vancomycin, are naturally more toxic than others, so they require routine blood-level monitoring to ensure that concentrations remain at effective, but not dangerous, levels. The cost of the assays is not insignificant, and it needs to be added to the cost of the drug itself when decisions are made on which antibiotics should be recommended in a hospital formulary for the treatment of particular infections.

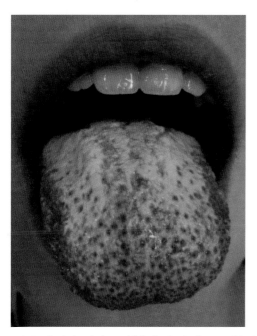

Figure 9.3   Oral thrush in a child who had taken antibiotics. *Source:* http://commons.wikimedia.org/wiki/File:Thrush.JPG.

In some cases the side effects of antibiotics are attributable to their activity on the normal flora of the body. Ampicillin, though orally active, is still not particularly well absorbed from the gastro-intestinal tract. This means a significant fraction of an oral dose reaches the colon and kills the 'friendly' bacteria there – organisms like *Lactobacillus* and *Bifidobacterium* species for example – with the effect that ampicillin is prone to causing diarrhoea on prolonged use; a similar situation arises with some oral cephalosporins and with clindamycin. Eradication of the normal flora may also predispose to overgrowth of organisms whose numbers are normally kept in check by competition with the resident species; both *Clostridium difficile* and *Candida albicans* may be present in low concentrations in the gastro-intestinal tract of healthy individuals, but exhibit a population explosion giving rise to the infections termed pseudomembranous colitis and thrush (Figure 9.3) respectively when the competition is removed. Fluoroquinolone antibiotics, in particular, predispose to *Cl. difficile* and MRSA colonization.

9.2.3 Oral absorption and dose frequency

Patient compliance is particularly important if uniform, effective blood concentrations are to be maintained, and noncompliance, which predisposes to treatment failure and resistance development, becomes a particular problem with infections requiring protracted or indefinite therapy such as tuberculosis, cystic fibrosis infections and HIV/AIDS. Compliance is promoted by the use of orally active antibiotics. Drugs that can only be given by injection are relatively little used in community medicine because it is inconvenient for patients to receive regular injections at home. This, in turn, means that the manufacturers have to recover their development costs through hospital sales alone, which naturally makes the antibiotics more expensive. The requirement to take antibiotic doses every six hours, which arises with some older penicillins and tetracyclines for example, is unpopular with patients because, if strictly applied, it would curtail sleep; for this reason more modern antibiotics have tended to exhibit longer half-lives leading to once- or twice-daily dosing.

9.2.4 Mode of action: bactericidal or bacteriostatic

Antibiotics are often described as either 'bactericidal' (meaning that they kill infecting bacteria at concentrations that can safely be achieved at the infection site) or 'bacteriostatic' (meaning that they merely inhibit growth without necessarily killing the bacteria) but this simple classification can obscure some important points:

- Antibiotic concentrations vary in different body fluids. Concentrations in the urine are often much higher than those in the blood, for example, whereas concentrations in cerebrospinal fluid and bile may be much lower. Thus a bactericidal concentration in the blood may correspond to one which is merely bacteriostatic at another infection site.
- Some antibiotics, such as the β-lactams, are only really effective against growing bacteria, and the slower the growth rate, the poorer the antibiotic action. Thus, bacteria that are immersed in a polymeric biofilm which may, in turn, be covered in mucus – in the respiratory tract for example – may be growing very slowly indeed, and though laboratory tests show the organism to be sensitive to a particular antibiotic, the reality is that the drug has very little effect.
- The rate at which bacteria are killed may, for some antibiotics, vary from one species to another. Vancomycin, for example, is only slowly bactericidal anyway and increasing its concentration produces only a small increase in the rate of kill; against *Enterococcus* species and *Staphylococcus haemolyticus* it is merely bacteriostatic.

Table 9.1 Some important antibiotics displaying concentration-dependent and time-dependent killing.

Concentration dependent	Time dependent	Mixed properties: both concentration and time influence outcome
Therapy should maximize the concentration that can be safely achieved	Therapy should maximize duration of exposure above MIC[a]	Therapy should maximize the amount of antibiotic that can be safely given
aminoglycosides	β-lactams	tetracyclines
fluoroquinolones	erythromycin, linezolid	vancomycin[b]
metronidazole		

[a]Minimum inhibitory concentration.
[b]Vancomycin is described in some textbooks as predominantly time dependent.

Despite these complicating factors it is, nevertheless, worth noting that antibiotics that are bactericidal, for example β-lactams, aminoglycosides, quinolones, glycopeptides and rifampicin, are normally preferred to bacteriostatic agents, such as tetracyclines, macrolides and sulphonamides, for the treatment of immunosuppressed patients simply because bacteriostatic antibiotics just stop the bacteria growing and so allow the body's immune system an opportunity to eradicate the infection. If the immune system is not functioning correctly there is a risk of relapse after the course of antibiotics is completed.

The bacterial killing effect of antibiotics may be influenced either by the concentration of drug to which the antibiotic is exposed or the length of exposure, or both. It is important, therefore, to know which of these is most likely to influence the outcome of treatment. Table 9.1 lists some of the major groups in each category.

Figure 9.4 shows the typical serum concentration profile following a dose of antibiotic. The peak concentration (Cmax) should be as high as possible for concentration-dependent antibiotics, whereas for time-dependent drugs the period of time for which the serum concentration exceeds the MIC is the most important factor. When both concentration and time influence the antibiotic efficacy the area under the curve (shaded red in Figure 9.4) should be as large as possible.

9.2.5 Tissue penetration

The ability to penetrate rapidly and deeply into the tissues and to accumulate inside host cells, particularly macrophages, is an important antibiotic attribute because a significant number of pathogenic bacteria reside at inaccessible sites in the body and survive, or worse, actually reproduce inside macrophages, which spread them round the body from the initial infection site. Many of the bacteria that are most adept at survival within macrophages are encountered relatively infrequently, such as chlamydia, *Legionella*, *Listeria* and *Brucella* species, but some, like *Mycobacterium tuberculosis*, are much more important from a worldwide perspective and yet others like *Salmonella* species and *Staphylococcus aureus* are relatively commonplace. Intracellular permeation varies substantially between the different families of antibiotics. Among the major antibiotic groups:

- β-lactams, vancomycin and aminoglycosides are not concentrated inside mammalian cells to any significant extent; data show varying pictures depending on experimental conditions but the intracellular concentration of these drugs is often lower than that in the extracellular fluid.
- Tetracyclines and, to an even greater extent, rifampicin, fluoroquinolones and macrolides (particularly azithromycin) are often reported to achieve concentrations inside macrophages that are many times higher than

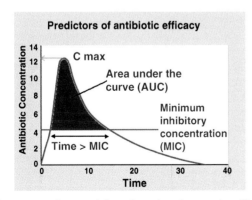

Figure 9.4 Factors influencing the therapeutic efficacy of an antibiotic.

those outside. Generally, these same drugs exhibit good tissue penetration too.

- Lipid-soluble antibiotics like trimethoprim are usually well dispersed in the tissues, exhibit long half lives and so require less frequent dosing.

Some antibiotics also exhibit a 'postantibiotic effect' in which bacterial growth continues to be suppressed even after administration of the drug has ceased and its plasma levels have fallen below the recognized minimum inhibitory concentration. Fluoroquinolones and macrolides possess an advantage over penicillins by exhibiting long postantibiotic effects, but the property is also seen in aminoglycosides, and again, it predisposes to less frequent dosing and is regarded as a good characteristic in an antibiotic.

9.2.6 Interactions with other drugs

An ideal antibiotic should not interact with other drugs, but unfortunately several common antibiotics do. Occasionally interactions arise between antibiotics themselves: β-lactams and aminoglycosides are weak acids and bases respectively, so they should not be present together in the same transfusion fluid because of a significant risk of diminished aminoglycoside activity. Interactions that either increase the activity or toxicity of nonantibiotic drugs are more commonly seen: macrolides, for example, increase the toxicity of digoxin, phenytoin, theophylline and warfarin, whereas rifampicin diminishes the anticoagulant activity of warfarin; the Merck Manual online (http://www.merckmanuals.com/professional/index.html) lists many other examples under the heading 'Common effects of antibiotics on other drugs'.

9.2.7 Resistance development

Antibiotics differ in their vulnerability to resistance development, particularly as a result of mutation. Although this would not necessarily be a major factor in the selection of an antibiotic, it may influence the duration of treatment or the decision on whether to use that antibiotic alone or in combination with another (see the following section). Rifampicin, for example, is rarely used on its own because resistance is acquired relatively easily, particularly by staphylococci and by some streptococci which may exhibit a mutation rate approximately 100 times greater than *Mycobacterium tuberculosis* (the other main organism against which

rifampicin is used). When the population of bacteria at the infection site is large (when the infection is severe and well established) high mutation rates mean that resistant organisms can readily arise during treatment.

9.3 Antibiotic combinations

In the early days of antibiotic therapy, combination products comprising two or more different antibiotics in fixed ratios were common, but that is no longer the case. The current philosophy is that the single most appropriate antibiotic should be used and those with narrow spectra of activity are to be preferred over broad spectrum agents in order to restrict antibiotic resistance development. It is now commonly accepted that the circumstances or justifications for using antibiotic combinations are restricted to the following:

- To provide broad-spectrum cover for a short time whilst the causative organism of a severe infection is being identified in the laboratory or, in the case of neutropenic patients, admitted to hospital with fever of unknown origin (neutropenic patients are those with low neutrophil counts who, consequently, are particularly vulnerable to infection).
- When there is a mixed infection (caused by two or more organisms) for which a single effective antibiotic is not available. The use of multiple antibiotics to reduce the microbial load in the gastro-intestinal tract prior to surgery is a similar situation; frequently metronidazole is used to kill anaerobic bacteria, together with another antibiotic such as a cephalosporin or aminoglycoside, although many different antibiotic 'cocktails' are used for this purpose.
- To restrict antibiotic resistance development, particularly in infections where the treatment is of long duration, for instance tuberculosis, lung infections in cystic fibrosis and HIV/AIDS. This is based on the principle that the chances of an organism mutating to become resistant to two dissimilar antibiotics simultaneously is very much lower than the probability of resistance development when a single agent is used.
- To achieve synergy (where the combined effect of two antibiotics is greater than the sum of their individual effects). Antibiotic synergy is relatively easy to demonstrate in the laboratory, but convincing evidence of synergy is less common *in vivo*, partly because it is often only evident at particular concentration ratios which

may not be easy to achieve at the infection site. It is widely accepted, however, that aminoglycosides and β-lactams display synergy in several infections, particularly endocarditis.

- Using two drugs together may minimize toxicity. There are, however, few circumstances when this strategy is put into practice; the most commonly quoted example being the combination of flucytosine with amphotericin for the treatment of severe fungal infections, which has been claimed to result in a lower incidence of kidney damage than that resulting from the use of amphotericin alone.

9.4 Using antibiotics for prophylaxis

Again, the trend in recent years has been to use antibiotics more circumspectly than was the case in the second half of the twentieth century, and the prescribing of prophylactic antibiotics is (or should be) now largely restricted to the following well defined circumstances.

To protect:

- Patients from infection arising from surgical procedures at sites in the body where bacteria are normally present, for example the colon, vagina and upper respiratory tract. Antibiotics used in this way are given for a short time prior to the surgery itself and not normally for more than 48 hours in total.
- Patients with heart-valve defects or prostheses from infection arising from surgical procedures elsewhere in the body (such as dental surgery) because bacteria carried in the blood are likely to colonize the heart.
- Healthy persons who are known to have been exposed to dangerous pathogens, such as organisms causing meningitis or anthrax. It is claimed that up to 50% of persons who live in temperate climates succumb to gastro-intestinal infections when travelling in tropical areas, and prophylactic antibiotics are sometimes used in this situation too.
- Immunocompromised patients, particularly HIV/AIDS patients and those receiving chemotherapy or radiotherapy, from opportunist infections, e.g. *Pneumocystis jiroveci* pneumonia.

Acknowledgement

Chapter title image: http://commons.wikimedia.org/wiki/File:Penicillin_cures_gonorrhea.jpg

Chapter 10

Antibacterial antibiotics

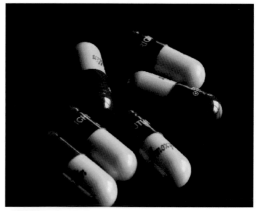

Amoxicillin capsules

KEY FACTS

- Penicillins account for 55% of UK health service general practice prescriptions for antibiotics, with macrolides, tetracyclines and trimethoprim each accounting for approximately 10%.
- Cephalosporins are grouped in four 'generations' with the trend of each successive generation possessing greater intrinsic activity towards Gram-negative species coupled with better enzyme resistance.
- The carbapenems are injection-only β-lactam antibiotics that have a broad spectrum of activity and are largely resistant to the extended-spectrum β-lactamases which inactivate third-generation cephalosporins.
- Of the macrolides, erythromycin has a spectrum largely limited to Gram-positive bacteria and suffers from erratic intestinal absorption and poor acid stability. These faults were corrected to varying degrees in the more recent macrolides.
- The tetracyclines are broad-spectrum, orally active antibiotics whose importance has diminished due to resistance development.
- Trimethoprim accounts for over 9% of UK general-practice prescriptions and is used largely for the treatment of urinary tract infections (UTI). Fluoroquinolones are also used for UTI as well as respiratory and other infections.
- Most of the antibiotics for tuberculosis are used exclusively for that purpose; the important ones are all orally active, which promotes compliance during the long course of treatment.
- Aminoglycosides and glycopeptides are not absorbed from the gastro-intestinal tract, so their injections are largely used for hospital treatment of dangerous Gram-negative and Gram-positive infections respectively. Both classes of antibiotic are relatively toxic and require blood-level monitoring.

Essential Microbiology for Pharmacy and Pharmaceutical Science, First Edition. Geoffrey Hanlon and Norman Hodges.
© 2013 John Wiley & Sons, Ltd. Published 2013 by John Wiley & Sons, Ltd.

The importance of antibiotics as a drug category is reflected in the fact that they regularly feature in the top six classes of drugs most frequently dispensed in the UK. Over 47 million health service prescriptions for community-use anti-infectives were dispensed in 2010 at a cost of approximately £222 million, and the great majority of these (85%) were for antibiotics to treat bacterial infections (drugs to treat protozoal, fungal and viral infections are discussed in Chapters 5, 11 and 12 respectively). Currently there are 69 antibacterial antibiotics described in the UK British National Formulary, so clearly there is not the scope within this chapter to consider them individually. One approach to categorizing this large variety of drugs is to group them by their mechanism of action. This may appeal to a microbiologist or biochemist but it does lead to some quite dissimilar groups of drugs arising under the same heading. For example, both the aminoglycosides and the macrolides act by inhibiting bacterial protein synthesis, but they differ in terms of their antibacterial spectra, oral absorption, side effects, uses and many other characteristics. Consequently, this chapter will mainly comprise a consideration of the major antibiotics grouped on the basis of their chemical structures because this largely determines their properties and the practical aspects of their use.

10.1 The frequency of use of the major groups of antibiotics

Figure 10.1 shows the proportions of antibiotics prescribed in the major therapeutic classes in the UK in 2010.

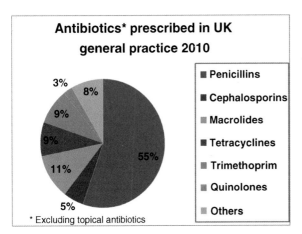

Figure 10.1 Antibiotics prescribed in UK general practice in 2010.

It is clear that, despite their age, the penicillins are, by far, the most frequently prescribed group of antibiotics, and penicillins and cephalosporins together account for 60% of all prescriptions. Trimethoprim is, like penicillins, very long established, and it too is still widely prescribed. Macrolides and tetracyclines are both more frequently used than cephalosporins, whereas quinolones, despite their rise in popularity over the last 30 years and the increasing number of drugs available in this class, still represent only about 3% of general practice prescriptions. It should be emphasized though that the relative popularity of the different groups would vary significantly from one country to another: penicillins would be relatively less popular and cephalosporins more popular in the United States for example, and the numbers would differ again if they were taken from hospital inpatient data where injectable antibiotics like glycopeptides and aminoglycosides would be more prominent.

The groups of antibiotic classes considered below are those listed in Figure 10.1 and, in each group, attention is focussed particularly on those drugs currently available in the UK. Many alternative antibiotics are available in other countries.

10.2 Penicillins

The penicillins (Table 10.1) were the first true antibiotics and have been in use since the 1940s. The early drugs, benzylpenicillin and phenoxymethylpenicillin, were susceptible to β-lactamases and were only active against Gram-positive species and Gram-negative cocci, but these limitations were overcome in the 1960s by the introduction of several semisynthetic penicillins – meticillin (formerly spelt methicillin), flucloxacillin, amoxicillin and others – manufactured by substituting new side chains (R in Figure 10.2) onto the 6-aminopenicillanic acid nucleus. Susceptibility to β-lactamases also became less of a problem with the development of molecules with greater intrinsic β-lactamase resistance (for example, temocillin) and the introduction of combination products comprising a penicillin with a β-lactamase inhibitor (for example, amoxicillin and ticarcillin, each with clavulanic acid as the inhibitor, and piperacillin with tazobactam). Research interest tended to shift from penicillins to cephalosporins after about 1980 and no new penicillins have been introduced since then. However, they are still the most heavily prescribed class of antibiotics in the UK and owe their continuing popularity to their low toxicity and side effects (mostly allergic reactions), their relative cheapness and the fact that, as a group, they cover most of the major bacterial pathogens. Meticillin was the first

Table 10.1 The characteristics of selected penicillins.

Penicillin	Characteristics
Benzylpenicillin[a] (penicillin G)	The first penicillin. Active against Gram-positive species and Gram-negative cocci; very susceptible to β-lactamase; still the first choice antibiotic for infections due to many streptococci, clostridia and (now rarely encountered) sensitive staphylococci.
Phenoxymethyl penicillin (penicillin V)	Orally active and less potent, but otherwise similar to benzylpenicillin. Sometimes used for respiratory infections in children, and for tonsillitis, but little used otherwise. One of the least expensive antibiotics.
Flucloxacillin	Its only, but important, use is for infections by β-lactamase-producing staphylococci (i.e. most staphylococcal infections); it is ineffective for MRSA.
Ampicillin	First broad-spectrum penicillin, active against the same organisms as benzylpenicillin and against many Gram-negative enterobacteria; it is β-lactamase-sensitive and incompletely absorbed after oral administration. Superseded by amoxicillin.
Amoxicillin	Similar antimicrobial spectrum to ampicillin but better absorbed so less frequent dosing. Combined, in the combination product co-amoxiclav, with the β-lactamase inhibitor clavulanic acid, which further extends the antimicrobial spectrum.
Ticarcillin[a]	A broad-spectrum penicillin, but used primarily for the treatment of serious Gram-negative infections including those by *Pseudomonas aeruginosa*. Not available on its own in the United Kingdom but only in combination with clavulanic acid.
Piperacillin[a]	Slightly more active against *Pseudomonas aeruginosa* and other Gram-negative species than ticarcillin but otherwise similar; widely available in combination with tazobactam as a β-lactamase inhibitor.
Temocillin[a]	Little activity against Gram-positive cocci, so used largely for infections by those Gram-negative species that produce β-lactamases – to which it is highly resistant. It has a long half-life permitting only twice-daily dosing.
Pivmecillinam	A pro-drug of mecillinam, it has little activity against Gram-positive species and is used largely for the treatment of urinary tract infections.

[a] = injection only.

β-lactamase-stable drug in the group and was superseded by flucloxacillin which, for many years, has been the antibiotic of choice for infections by β lactamase-producing *Staphylococcus aureus*. The rise in incidence of meticillin-resistant *Staphylococcus aureus* (MRSA), which, in some countries, accounts for 30% or more of staphylococcal bloodstream infections, is perhaps the most problematic gap in the spectrum because it normally necessitates the use of more toxic antibiotics like vancomycin.

Penicillins are susceptible to inactivation by hydrolysis of the β-lactam ring (Figure 10.2). This is the reaction accelerated by bacterial β-lactamases, but the hydrolysis occurs spontaneously anyway so penicillins cannot be stored for long periods as aqueous solutions. Injections are manufactured as freeze-dried powders to be reconstituted with water for administration, and oral syrups are dried granules which, after dissolving in water, can only be stored at refrigeration temperatures for no more than 7-14 days. Table 10.1 shows the characteristics of the penicillins available in the UK.

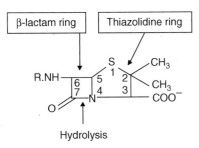

Figure 10.2 Ring numbering and point of hydrolysis in the penicillin nucleus.

Table 10.2   Characteristics of selected cephalosporins.

Cephalosporin and (generation)	Comments
Cefradine (1) Cefalexin (1) Cefadroxil (1)	Inexpensive, orally active drugs resistant to staphylococcal β-lactamases but susceptible to Gram-negative enzymes. Used for the treatment of uncomplicated urinary-tract infections and respiratory tract infections, but of declining use.
Cefaclor (2) Cefuroxime (2)	Cefaclor is orally active and similar to the first-generation drugs but possesses better activity against *Haemophilus influenzae*, as does cefuroxime which is usually injected, though an orally active cefuroxime axetil ester is available.
Cefixime (3) Cefpodoxime (3) Cefotaxime (3) Ceftazidime (3) Ceftriaxone (3)	Cefixime and cefpodoxime are orally active and have largely superseded the first generation drugs; cefixime has a longer duration of action than other cephalosporins and cefpodoxime has good activity against respiratory pathogens. The last three are broad-spectrum, injection-only antibiotics which tend to be used more for the treatment of serious Gram-negative infections. Ceftazidime is the only one with activity against *Pseudomonas aeruginosa* and ceftriaxone has a long plasma half-life, which means less frequent dosing.
Cefpirome (4)	Possesses most of the attributes of the injectable third generation drugs, including activity against *Pseudomonas aeruginosa,* and with greater β-lactamase stability.

10.3 Cephalosporins

Cephalosporins are similar to penicillins in many respects: they have similar structures, uses and mechanisms of action, and both groups possess the β-lactam ring in the molecule. The cephalosporins possess a six-membered dihydrothiazine ring rather than a thiazolidine ring linked to the β-lactam and this provides more opportunities for chemical substitution and the creation of semisynthetic cephalosporins. The early cephalosporins produced in the 1960s were largely unaffected by staphylococcal β-lactamases and so they were considered to be more reliable than penicillins for the treatment of staphylococcal infections. These so-called first-generation cephalosporins also possessed better activity against Gram-negative species than the early penicillins and so they were, and still are, used for the treatment of urinary-tract infections due to organisms like *E. coli*, *Proteus* and *Klebsiella* species, but they are susceptible to Gram-negative β-lactamases so bacteria producing these enzymes are likely to be resistant. Successive generations of cephalosporins have been produced since at least the 1970s (Table 10.2) and the trend has been for each successive generation to possess greater intrinsic activity towards Gram-negative species coupled with better enzyme resistance, but this has, in most cases, been accompanied with reduced activity towards Gram-positive species.

10.4 Other β-lactam antibiotics

One strategy to protect penicillins against enzyme inactivation is to combine them with a β-lactamase inhibitor. Clavulanic acid was the first such molecule to be developed and it is, itself, a β-lactam antibiotic, although it possesses such weak activity that it could not be marketed as an antibiotic in its own right. Clavulanic acid was first used in the early 1980s to protect amoxicillin against both staphylococcal and Gram-negative β-lactamases and the combination product co-amoxiclav consequently displays good antimicrobial activity against strains of *Staphylococcus, Klebsiella* and several other Gram-negative species that would otherwise be resistant. Clavulanic acid is used in a similar manner to protect ticarcillin in the combination product Timentin™, and both sulbactam and tazobactem (which are also β-lactams) have been combined with ampicillin and piperacillin respectively to achieve the same effect.

More recently, research attention has moved towards the carbapenems, which are structurally similar to penicillins but with the sulfur atom of the thiazolidine ring replaced by carbon. This group of β-lactams consists of the injectable antibiotics imipenem, meropenem, ertapenem and doripenem, which owe their value to their broad spectrum of activity and, particularly, their resistance to the extended spectrum β-lactamases (ESBLs). These enzymes have caused much concern in recent years by enabling an increasing number of strains of Gram-negative species to display resistance to the third-generation cephalosporins. The carbapenems, therefore, are the drugs of choice and the last line of defence against these dangerous pathogens. Imipenem is susceptible to inactivation by human renal dipeptidase and is only available in combination with the enzyme inhibitor cilastatin. The other three carbapenems do not exhibit this weakness and are available on their own. Extended spectrum β-lactamases also inactivate aztreonam, which has similar side chains and is used in similar circumstances to the third-generation cephalosporins, but differs from them in not having a second ring fused to the β-lactam ring in the nucleus of the molecule.

10.5 Macrolides

Erythromycin, discovered in 1952, was the first and, arguably, still the most important, macrolide, although it suffers from several disadvantages that prompted the development of semisynthetic macrolides like clarithromycin, azithromycin and, more recently, telithromycin. Strictly speaking, azithromycin and telithromycin are an azalide and a ketolide respectively but they are commonly referred to as macrolides, as they will be here.

Erythromycin, which possesses a similar antibacterial spectrum to the early penicillins, became useful as an alternative for penicillin-allergic patients and for the treatment of infections by β-lactamase-producing staphylococci, such as boils and impetigo (Figure 10.3).

Its limited spectrum (confined largely to Gram-positive species), poor acid stability, erratic oral absorption with relatively common gastrointestinal side effects, and its vulnerability to resistance development makes erythromycin far from an ideal antibiotic, although it is generally regarded as safe. Clarithromycin is slightly more potent, has better stability, absorption and tissue distribution, all of which lead to less frequent dosing (twice daily rather than four times with erythromycin), but its antibacterial spectrum is similar. Both drugs are used to treat oral and respiratory infections (including whooping cough and legionnaire's disease), as well as infections by chlamydia, mycoplasmas and spirochaetes (for example, Lyme disease). Azithromycin on the other hand has rather less activity against Gram-positive species than erythromycin but it possesses significant more activity against Gram-negatives, particularly enteric bacteria, and displays even better tissue distribution than clarithromycin. It is the drug of choice for sexually

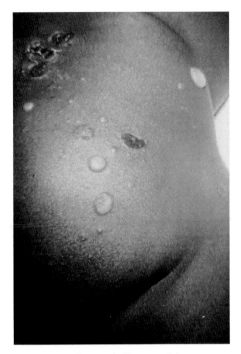

Figure 10.3 Impetigo typically caused by streptococci or staphylococci. *Source:* PHIL ID #5154; Centers for Disease Control and Prevention.

transmitted chlamydial infections, for which it can be obtained without a prescription in the United Kingdom.

Telithromycin is active against some species that have acquired resistance to other macrolides including penicillin- and erythromycin-resistant *Streptococcus pneumoniae*. The UK advice is that it should be reserved for respiratory infections caused by organisms resistant to β-lactams and other macrolides.

10.6 Tetracyclines

The tetracyclines are a group of orally active, bacteriostatic, broad-spectrum antibiotics, which, in the late 1950s, were the largest selling antibiotics in the world, but they have since declined in importance due mainly to the development of resistance. Tetracycline itself, oxytetracycline and chlortetracycline were developed in the 1940s and 1950s, at which time they were widely used for the treatment of a variety of Gram-positive and Gram-negative infections, but their value as an alternative to penicillins for penicillin-allergic patients diminished as resistance in *Staphylococcus aureus* became widespread.

The more recently developed minocycline and doxycycline are more potent, lipophilic and better absorbed from the gastrointestinal tract than the early drugs, so they

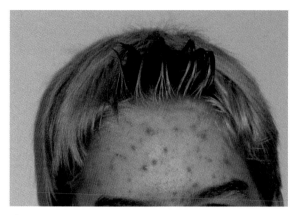

Figure 10.4   Acne in a 14-year-old male. Acne is typically caused by *Propionibacterium acnes* and may be treatable with tetracyclines or erythromycin. *Source*: http://upload.wikimedia.org/wikipedia/commons/4/4b/Akne-jugend.jpg.

achieve effective concentrations in a wider range of tissues and require less frequent dosing. Currently the tetracyclines are the drugs of choice for the treatment of infections by some relatively rarely encountered groups of organisms like chlamydia (for example, trachoma), rickettsia (for example, typhus and Q-fever) and spirochaetes (for example, Lyme disease); they are also used in for the treatment of oral infections, acne (Figure 10.4) and some respiratory infections; doxycycline is used in malaria prophylaxis.

The absorption of tetracyclines is inhibited by divalent and trivalent cations, so they should not be taken with milk, antacids and iron-containing products. They are yellow in colour and because of their affinity for calcium they accumulate in developing bones and stain the teeth of unborn babies and young children, for whom they are contraindicated.

Interest in semisynthetic tetracyclines has been revived by the introduction, in 2005, of tigecycline (a glycylglycine), which shares many of the properties of tetracyclines but possesses activity against several organisms that are resistant to them, such as MRSA and vancomycin-resistant enterococci. Tigecycline is administered by injection and it is recommended that it be reserved for the treatment of complicated skin and soft-tissue infections and complicated abdominal infections caused by multiple-antibacterial resistant organisms.

10.7 Sulphonamides and trimethoprim

Sulphonamides are orally active, bacteriostatic drugs, many of which were developed during the early and mid-twentieth century but whose use has diminished substantially since the 1970s due both to resistance and to side effects which rendered them less suitable than many of the β-lactam, quinolone and other antibiotics. They are now rarely prescribed on their own for systemic action although they are still used in topical products and in combination with trimethoprim. Currently, sulfadiazine is used for the prevention of rheumatic fever and the treatment of toxoplasmosis, and sulfamethoxazole, now the only other systemically acting sulfonamide, is more important because of its synergistic combination with trimethoprim (co-trimoxazole), which is the drug of choice for *Pneumocystis jiroveci* pneumonia (to which immunocompromised individuals are particularly susceptible) and is also used for *Nocardia* and *Toxoplasma* infections.

Trimethoprim, on its own, is widely prescribed for the treatment of uncomplicated urinary tract infections where it is effective in 80% or more of infections; it is also used in respiratory infections and for gastrointestinal infections due to *Shigella* (dysentery) and *Salmonella* species. It has fewer and less severe side effects than co-trimoxazole and sulphonamides, and is orally active and inexpensive.

10.8 Quinolones

The quinolones are synthetic, orally active, bactericidal antibiotics, the earliest of which, nalidixic acid, dates back to the 1960s. As with the cephalosporins, the quinolones have been grouped into generations, but again, there is no universal agreement on the generation to which each belongs (Table 10.3). The first generation – also referred to as Group 1 – differ from the later ones in that there is no fluorine atom in the molecule, so they are simply referred to as quinolones, and groups 2–4 are, or should be, distinguished as *fluoro*quinolones.

Fluoroquinolones became the most widely prescribed antibacterial agents for adults in the United States in 2002 but there have been increasing reports both of their widespread misuse to treat infections for which they are unlicensed or not recommended, and of significant side effects, which have resulted in several drugs in the class being withdrawn from clinical use. Two of the more prominent adverse effects are tendon damage (which precludes use of fluoroquinolones for children) and prolongation of the QT interval which is associated with cardiac arrhythmias.

Table 10.3 Uses of selected fluoroquinolones.

Quinolone (generation)	Routes and (typical doses/day)	Antibacterial spectrum and uses
Nalidixic acid (1)	Oral (4)	An antibacterial spectrum that is largely confined to *E. coli* and other Gram-negative enteric bacteria. Largely restricted to urinary-tract infections; still available but little used.
Norfloxacin (2)	Oral (2)	Used only for urinary-tract infections and chronic prostatitis.
Ciprofloxacin (2)	Oral and IV (2)	Has good activity against Gram-negative species (including *Pseudomonas aeruginosa*) but only moderate activity against Gram-positives; inactive against MRSA. Used primarily for urinary, respiratory (but not pneumococcal pneumonia), gastrointestinal and bone and joint infections.
Ofloxacin (2)	Oral and IV (1 or, more commonly 2)	Less active than ciprofloxacin against Gram-negative species, but levofloxacin particularly is more active against Gram-positives (again excluding MRSA). Both are used primarily for urinary and lower respiratory infections, as well as gonorrhoea and genital infections. Ofloxacin is a racemic mixture containing 50% of the L-isomer levofloxacin; the latter is more potent and so has the additional use as a second line treatment for community-acquired pneumonia.
Levofloxacin (2)[a]	Oral and IV (1–2)	
Moxifloxacin (4)	Oral (1)	More active against anaerobes and Gram-positive species than ciprofloxacin. Used for sinusitis, community-acquired pneumonia or exacerbations of chronic bronchitis.

[a]Sometimes classified as third generation.

10.9 Aminoglycosides

All of the aminoglycosides are bactericidal, water-soluble drugs that are only weakly active against anaerobic organisms and not absorbed from the gastro-intestinal tract. They are administered by injection for systemic activity. Gentamicin, the most important aminoglycoside in the UK, is used for the treatment of serious blood, urinary and respiratory infections, most commonly those by Gram-negative species, though occasionally for staphylococci too. Both gentamicin and its less commonly used alternative, amikacin, have the potential to cause ear and kidney damage, and their blood levels are routinely monitored during therapy of patients with impaired renal function.

In addition to its use in eye drops and many topical products, neomycin may be given orally to reduce the gut flora prior to surgery but it is too toxic to be used for a systemic effect and the orally administered drug is not absorbed into the systemic circulation. Tobramycin is nebulized for the treatment of chronic *Pseudomonas* *aeruginosa* infections in cystic fibrosis patients and streptomycin is now a relatively rarely used drug which is a second line option for drug-resistant tuberculosis.

10.10 Glycopeptides

The glycopeptides (vancomycin and teicoplanin) are molecules that are too large to pass through the porins of the Gram-negative outer membrane so their activity is confined almost exclusively to Gram-positive species. They are used by injection to treat serious infections, including those by *Enterococcus* species and MRSA. Like the aminoglycosides, they are not absorbed from the gut and they have the potential to cause both ear and kidney damage (particularly vancomycin; teicoplanin is less toxic) so blood-level monitoring may be required. Vancomycin is given orally to treat pseudomembranous colitis due to *Clostridium difficile* and, like neomycin (above), its absorption from the gut is so poor that the antibiotic action is localized there.

Table 10.4   The uses and characteristics of other selected antibiotics.

Name	Route[*]	Uses and characteristics
Rifampicin	O, I/V	Used in combination with other antibiotics for serious staphylococcal infections and with the three below for tuberculosis.
Isoniazid	O, I/V	All three used exclusively for tuberculosis, typically as a combination of three or four antibiotics (including rifampicin) for two months' intensive treatment, which is followed by four months of rifampicin and isoniazid alone.
Pyrizinamide	O	
Ethambutol	O	
Metronidazole	O, T, I/V	Used in a variety of formulations for anaerobic infections by bacteria and protozoa.
Linezolid	O	Complicated skin and soft-tissue infections caused by Gram-positive bacteria including vancomycin-resistant enterococci and MRSA. Regarded as drugs that should be reserved for this purpose.
Daptomycin	I/V	
Quinupristin/dalfopristin	I/V	

[*]O = oral; I/V = intra-venous; T = topical.

Teicoplanin affords several advantages over vancomycin, including the following:

- Its fatty acid side chains make the molecule 50–100 times more lipophilic than vancomycin; this gives a longer half life and better tissue penetration.
- Teicoplanin is more potent: its minimum inhibitory concentrations are 1/2 to 1/4 those of vancomycin.
- It is more acidic than vancomycin, so it forms a water-soluble sodium salt, which may be given intramuscularly.
- Teicoplanin achieves better intraleukocyte killing than vancomycin.

10.11   Other antibiotics

Most of the antibiotics used for the treatment of tuberculosis (TB) (Table 10.4) are employed exclusively for that purpose; the only exception is rifampicin, which is sometimes used for staphylococcal infections. Because of the long duration of TB therapy – six months or more – there is ample opportunity for the infecting bacteria to develop resistance so it is invariably the case that antibiotics for treating this disease are used in combinations of at least two and, in the initial phase, of three or four. Rifampicin and isoniazid are almost always components of the regimen; if the organism is resistant to both of these it is designated multiply resistant *Mycobacterium tuberculosis* (MRTB) and second line drugs, such as streptomycin, cycloserine and capreomycin, are called into play.

Metronidazole is a synthetic drug that is an alternative to vancomycin for *Clostridium difficile* infections and it is unusual in that it acts against any kind of cell growing anaerobically. Consequently, it can be used to treat infections by grossly dissimilar organisms like bacteria, protozoa and even helminths. It may be formulated as oral, injectable, topical or rectal products, which, variously, have applications in the treatment of leg ulcers, pressure sores, bacterial vaginosis, pelvic inflammatory disease, gingivitis and surgical prophylaxis.

The increasing incidence of MRSA and the emergence of enterococci and staphylococci resistant in varying degrees to vancomycin and teicoplanin led to pharmaceutical companies developing antibiotics specifically to be used against these organisms. Three such drugs came into use between 1999 and 2003: linezolid, daptomycin and the synergistic combination product quinupristin/dalfopristin that is more commonly known by its proprietary name Synercid™. They are relatively expensive because of both their limited application and the official guidance that they should be regarded as valuable reserve drugs of last resort.

Acknowledgement

Chapter title image: http://commons.wikimedia.org/wiki/File:Medication_amoxycillin_capsule.JPG.

Chapter 11
Antifungal agents

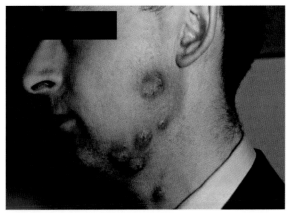

Ringworm which, despite its name, is a fungal infection

KEY FACTS

- The majority of fungi are harmless saprophytes and only very few are capable of causing disease in humans.
- Those that do cause infections can lead to very persistent disease, which may be difficult to eradicate and can be life threatening.
- Since fungi are eukaryotic cells they contain similar features to human cells. As a consequence, antifungal agents used to treat these infections frequently give rise to toxic effects.
- Most of the antifungal agents in common use target the fungal membrane.

There are an estimated 100 000 species of fungi and the vast majority of these are saprophytes, living on dead and decaying matter. Some (including *Candida albicans*) form part of the normal microbial flora of the skin and GI tract but only a few (approximately 100) are pathogenic to humans. However, when fungi do cause infections they are usually persistent and difficult to treat.

Fungi can cause illness in a variety of ways:

- Poisoning
  - ingestion of poisonous fungi, such as *Amanita phalloides* (death cap fungus); production of mycotoxins on foodstuffs.
- Inhaled allergens
  - during the summer months spores of *Alternaria* and *Cladosporium* can cause allergic asthma in susceptible patients.
- Infection
  - systemic;

- mucocutaneous;
- subcutaneous;
- cutaneous;
- superficial.

11.1 Common fungal infections

More details of fungal infections can be found in Chapter 4 and what follows here simply outlines the different infections in terms of their severity.

11.1.1 Systemic mycoses

Systemic mycoses involve the internal organs of the body and the fungus can become widely disseminated. Prior to the introduction of effective antifungal chemotherapy

Essential Microbiology for Pharmacy and Pharmaceutical Science, First Edition. Geoffrey Hanlon and Norman Hodges.
© 2013 John Wiley & Sons, Ltd. Published 2013 by John Wiley & Sons, Ltd.

these disseminated mycoses were invariably fatal and even now they represent a major challenge. The problem is exacerbated by the increased number of immunosuppressed patients now seen in hospitals.

Invasive fungal infections are important causes of illness in those patients:

- with neutropenia (shortage of neutrophils in the blood);
- who receive chemotherapy;
- who undergo haematopoietic stem-cell transplantation;
- with AIDS.

Persistent fever in patients with neutropenia receiving broad-spectrum antibiotics may be the only clinical indication of an invasive fungal infection. The incidence of such infections has increased over the past two decades and the trend involves not only severely compromised hosts but also noncompromised patients such as those on surgical and medical intensive care units, burns and neonatal units. Intensive care treatment is an independent risk factor for fungal infection.

- *Candida* species represent the fourth commonest cause of nosocomial (or hospital acquired) bloodstream infection, with an estimated 38% mortality.
- *Candida albicans* is the species responsible for over 80% of cases.
- The most commonly encountered agents of systemic mycoses in the UK are *Candida albicans*, *Aspergillus*, *Cryptococcus* and *Pneumocystis jiroveci* (formerly *Pneumocystis carinii*).

11.1.2 Mucocutaneous mycoses (Candidiasis)

Thrush – oral candidiasis takes the form of discrete or confluent white patches on the mucous membranes of the mouth and pharynx (Figure 9.3). It caused significant infant mortality in the nineteenth century and, even today, is common in young babies and in those patients suffering from immunosuppression. Up to 75% of women have reported at least one episode of vulvo-vaginal candidiasis and 40–50% have had two or more episodes.

11.1.3 Subcutaneous mycoses

These infections are usually initiated by penetration of the skin with contaminated splinters, thorns or soil. They are persistent infections common among farmers in hot climates. Examples include:

- Sporotrichosis
  Ulcerated lesion at the site of infection and multiple nodules and abscesses along draining lymphatics.
- Chromomycosis
  Warty, ulcerating, cauliflower-like growths.
- Maduromycosis
  Localized, destructive, granulomatous and pus-filled lesions.

11.1.4 Cutaneous mycoses

Two common examples of cutaneous mycoses are given below:

- Ringworm
  – Dermatophyte infection of the skin, nails and hair.
- Intertriginous candidiasis
  – Infection of skin areas which are moist for prolonged periods (nappy rash or incontinent, elderly patients).
  – Common in skin folds of obese patients.

11.1.5 Superficial mycoses

These are mainly cosmetic infections limited to uppermost layers of skin and hair and include:

- Pityriasis versicolor
  – Widespread, but most common in warm climates.
  – Asymptomatic white or tanned areas of skin on the trunk.
- Tinea nigra
  – Lesions confined to the palms, irregular, flat, darkly coloured areas.
  – Prevalent in tropics.
- White piedra
  – Grows on scalp or beard.
  – Soft pale nodules on hair shafts.
- Black piedra
  – Hard, dark nodules on infected hair shafts.

11.2 Agents used to treat fungal infections

An ideal antifungal agent should eradicate the infecting organism without affecting the host or the resident microflora. However, few agents achieve this and the result is usually a compromise.

Table 11.1 summarizes the different agents available to treat fungal infections, and the remainder of this chapter will provide more details of their characteristics.

A major problem when tackling fungal infections is that fungi and mammalian cells share characteristics at a biochemical level and so chemical agents that attack fungal targets also affect the host cell. Probably the greatest difference between the two cell types is in the structure of the cell wall but interestingly none of the older antifungal agents used this as a target and most act upon the fungal membrane or nucleus. There are now newer agents which act on fungal cell walls. Figure 11.1 summarizes the different modes of action of the various antifungal agents described in this chapter.

11.2.1 Polyene macrolides

These are antibiotics produced by bacteria belonging to the genus *Streptomyces*:

- Amphotericin B – *Streptomyces nodosus*
- Nystatin – *Streptomyces norsei*

Table 11.1 Agents used to treat fungal infections.

Class	Examples	Spectrum of activity	Mode of action
Polyene macrolides	Amphotericin Nystatin	Broad spectrum *Candida albicans*	Bind to membrane ergosterols affecting membrane fluidity.
Fluorinated pyrimidines	Flucytosine	*Candida albicans* *Cryptococcus sp.* *Torulopsis sp.*	Inhibits DNA biosynthesis and becomes incorporated into fungal RNA causing inhibition of protein synthesis.
Allylamines	Terbinafine Amorolfine	Dermatophyte fungi	Inhibit ergosterol synthesis by inhibition of the enzyme squalene epoxidase.
Antifungal antibiotics	Griseofulvin	Dermatophyte fungi	Inhibits cell division by interfering with microtubule formation in the nucleus.
Imidazoles	Clotrimazole Miconazole Ketoconazole Econazole Isoconazole	*Candida albicans* Broad spectrum Broad spectrum *Candida albicans* *Candida albicans*	Inhibit ergosterol synthesis by inhibition of cytochrome P-450 dependent 14α demethylation of lanosterol.
Triazoles	Fluconazole Itraconazole Posaconazole Voriconazole	*Candida albicans* *Cryptococcus sp.* Dermatophyte fungi *Candida albicans* Dermatophyte fungi *Pityrosporum* sp.	Same as imidazoles.
Echinocandins	Caspofungin Anidulafungin Micafungin	*Candida albicans* *Aspergillus sp.*	Block fungal cell wall synthesis by non-competitive inhibition of β(1,3) D-glucan synthase.
Sordarins	Sordarin	*Candida albicans* *Candida tropicalis* *Cryptococcus neoformans* *Pneumocystis jiroveci*	Potent inhibitors of fungal protein synthesis.

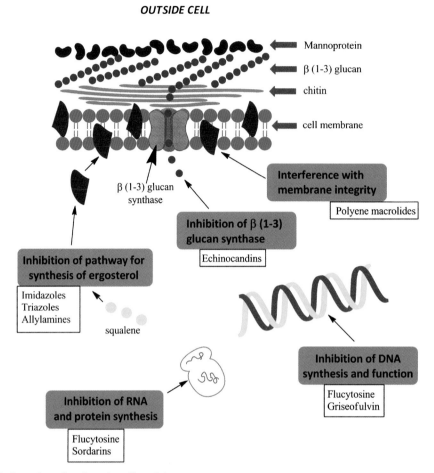

Figure 11.1   Various sites of action of antifungal drugs.

Both are toxic drugs; however amphotericin can be used systemically although nystatin is too toxic for other than topical administration. Nystatin is not absorbed from the gut after oral administration and so can be safely used for gastrointestinal infections. It is also used for *Candida* infections of the skin and mucosa. Amphotericin is absorbed to a limited extent but not enough to be used for oral administration. It is given parenterally and is invariably associated with side effects. Hydrophobic interactions occur between the ergosterol present in fungal membranes and the polyene macrolides leading to enhanced membrane permeability and leakage of intracellular ions. Polyenes are selective for ergosterol but have some affinity for cholesterol (found in mammalian cells), and this accounts for their toxicity. Formulations of amphotericin encapsulated in liposomes are available and this has been shown to reduce toxic effects. Amphotericin remains an important antibiotic used for serious fungal infections.

11.2.2 Flucytosine

This is a base analogue which is taken up into the fungal cell by the same transport system as adenine and cytosine. Once inside the cell it is rapidly deaminated to 5-fluorouracil by cytosine aminohydrolase (an enzyme only found in yeasts) hence its spectrum of activity is confined to yeasts (*Candida*, *Torulopsis* and *Cryptococcus*). The drug is then utilized in the formation of both DNA and RNA but as it is an analogue it results in malfunctioning nucleic acids (see Figure 11.2).

It is used intravenously in severe systemic yeast infections usually in combination with amphotericin or fluconazole, because resistance rapidly arises when used

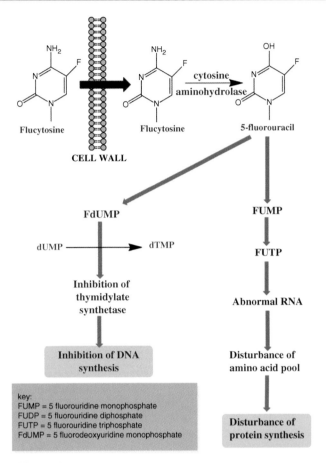

Figure 11.2   Mode of action of flucytosine.

alone due to a decrease in the activity of phosphorylating enzymes giving rise to FUMP and FdUMP.

11.2.3 Allylamines

Allylamines inhibit the enzyme squalene epoxidase, which is involved in the conversion of squalene to squalene epoxide, an important step in the synthesis of ergosterol (see Figure 11.3). Allylamines such as terbinafine have the following features:

- They are used orally and topically in the treatment of dermatophyte infections of the skin and nails.
- They are highly lipophilic and keratophilic drugs, which distribute in high concentrations in the stratum corneum, sebum, hair and nail matrix and persist for months after discontinuing therapy.
- Fingernail infections require 250 mg daily for 12 weeks; re-evaluated 18 weeks later. Toenail infections require 250 mg daily for 12 weeks; re-evaluated after 6–9 months. Longer courses are not usually more effective.

- Optimal clinical effect is not seen until new nail grows through.

11.2.3.1   Side effects

The British National Formulary currently lists over 25 different and significant side effects for terbinafine and hence it is not surprising that clinicians are reluctant to use this drug systemically.

11.2.4 Griseofulvin

Griseofulvin is a fungistatic antibiotic produced by the fungus *Penicillium griseofulvum*. It has been around for a long time and its use has diminished in recent years. The main characteristics of griseofulvin are as follows:

- It is orally active and accumulates in skin, nails, hair, fat and skeletal muscles.
- It is used in the treatment of ringworm of skin, nails and hair caused by dermatophyte fungi only.

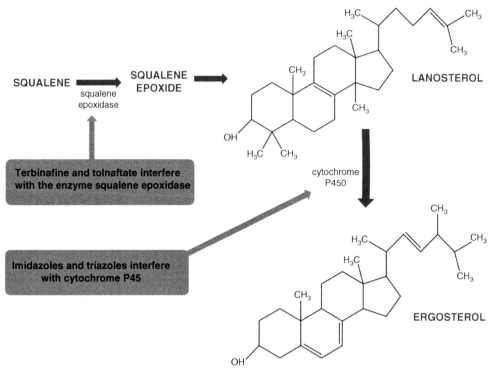

Figure 11.3 Synthesis of ergosterol indicating the points at which various antifungal agents interfere. Note that this synthetic pathway is not complete but merely indicates the stages at which interactions occur.

- The drug is lost from keratinized areas within days of stopping therapy.
- It requires a minimum of 2–4 weeks of treatment for tinea corporis and a year or longer for treatment of nail infections.
- Adsorption of drug from the gut is erratic and dependent on particle size.
- It is highly insoluble, and particles of 5 μm are required for formulation into tablets (ultra-micronized particles – less than 1 μm – are even better).

Griseofulvin works by disrupting the cell's mitotic spindle, thus arresting the fungal cells in metaphase. It also affects the structure of microtubules which are part of the cytoskeleton responsible for cell shape. They are made up of strong, rigid protein fibres packed into bundles and comprise a protein scaffold which provides support, internal organization and even movement of cells. They also help cells resist external mechanical stress.

11.2.5 Azoles

This is a generic name to describe the imidazole and triazole groups of antifungal agents. These possess within their structure a five-membered ring with either two (imidazoles) or three nitrogen atoms (triazoles).

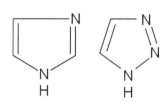

There are a number of both imidazoles and triazoles and it is not appropriate here to give details of each of them. Below is a summary of the characteristics of the most commonly used drugs.

Clotrimazole
- A topical agent used as a cream, lotion, solution, vaginal tablets and lozenges.
- Effective against a wide range of microorganisms.
- Used to treat candidiasis, dermatomycoses and superficial mycoses.

Miconazole
- Used for the prevention and treatment of oral and intestinal fungal infections.
- Not used systemically – can cause nausea, vomiting and diarrhoea. Isolated reports of hepatitis (care in hepatic impairment).

- Various topical formulations also available for the treatment of dermatophyte infections.

Ketoconazole
- Orally active – it can also be used topically.
- Dermatophyte infections of skin and finger nails.
- Indicated for systemic mycoses:
  - serious chronic mucocutaneous candidiasis;
  - serious gastro-intestinal mycoses;
  - prophylaxis of mycoses in immunosuppressed patients.
- Ketoconazole is better absorbed by mouth than other imidazoles.
- It has been associated with fatal hepatotoxicity;
  - the Committee on Safety of Medicines has advised that prescribers should weigh the potential benefits of ketoconazole treatment against the risk of liver damage.
- Need to monitor patients carefully both clinically and biochemically if this is given for longer than 14 days.

Fluconazole
- Orally active against yeasts and dermatophytes but no antibacterial activity.
- Used to treat candidiasis and cryptococcosis.
- Can cause liver and/or renal impairment.

Itraconazole
- Orally active but poor bioavailability.
- Spectrum of activity similar to fluconazole, but has significant activity against *Aspergillus.*
- Used to treat candidiasis, histoplasmosis, aspergillosis and unresponsive dermatophyte infections.

Voriconazole
- Broad spectrum of activity against yeasts and moulds, including fluconazole-resistant strains.
- Available in both IV and oral formulations.
- Excellent bioavailability.
- Low incidence of side effects.
- Licensed for use as primary treatment of invasive aspergillosis.

Posaconazole
- Similar spectrum to voriconazole.
- Available in oral formulation only.
- Used for the treatment of invasive aspergillosis either unresponsive to, or in patients intolerant of, amphotericin or itraconazole.

The imidazoles and triazoles work by inhibiting the formation of ergosterol which as we have said before is a vital component in fungal cell membranes (see Figure 11.3). They inhibit the enzyme cytochrome P450 which is responsible for the 14α-demethylation of lanosterol. This leads to a buildup of lanosterol and a depletion of ergosterol. It should be noted that cytochrome P450 is one of a family of liver enzymes responsible for the metabolism of a range of other drugs. Inhibition of this enzyme in the patient's liver may well therefore interfere with the metabolism of any other medicines they may be taking and so this should always be checked.

11.2.6 Echinocandins

The echinocandins are a relatively new class of antifungal drugs which block fungal cell-wall synthesis. They are semisynthetic lipopeptides produced by various species of fungi and inhibit the synthesis of β(1-3) D-glucan; a target unique to the fungal cell. They are currently only available in IV formulations but caspofungin possesses good activity against *Candida* and *Aspergillus* species; *Cryptococcus neoformans*, however, is resistant. Their safety profile is good and unlike the azole group, their prospects for drug interactions are low, as is the emergence of resistance.

11.2.7 Sordarins

These are another relatively new group of agents which are inhibitors of fungal protein synthesis by interfering with the action of an enzyme called elongation factor 2. This factor appears to be unique to yeasts and is responsible for the translocation of mRNA across the ribosomal reading frame during protein synthesis. The sordarins can be given by injection or orally and their spectrum of activity includes *Candida* species, *Cryptococcus neoformans* and *Pneumocystis jiroveci*.

11.3 Treatment of common fungal infections

The following is a summary of the BNF recommendations for the management of fungal infections.

Aspergillosis
- Voriconazole – drug of choice.
- Amphotericin or itraconazole if this fails.
- Caspofungin is licensed for aspergillosis unresponsive to amphotericin or itraconazole.

Candidiasis

- *Superficial infections of skin* treated topically (clotrimazole).
- *Oropharyngeal* – treat topically; give fluconazole by mouth if unresponsive. Use itraconazole if unresponsive to fluconazole.

Deep and disseminated candidiasis

- Amphotericin used alone or with flucytosine. Alternative: fluconazole.
- Voriconazole licensed for fluconazole-resistant candidiasis.
- Caspofungin used in fluconazole-resistant candidiasis unresponsive to amphotericin.

Cryptococcosis

- *Meningitis* – amphotericin with flucytosine IV for 2 weeks, followed by fluconazole by mouth for 8 weeks.
- Fluconazole given alone is an alternative in AIDS patients with cryptococcosis (not meningitis).

Histoplasmosis

- Itraconazole is used for immunocompetent patients and for prophylaxis.
- Ketoconazole is an alternative treatment.
- Amphotericin is preferred in severe infections.

Skin and nail infections

- Mild skin infections (dermatophyte) usually respond to local therapy (miconazole).
- Itraconazole, fluconazole or terbinafine if topical therapy fails.
- Terbinafine is the drug of choice for nail infections. Itraconazole may be used as intermittent therapy.

Acknowledgement

Chapter title image: PHIL ID #4807; Centers for Disease Control and Prevention.

Chapter 12
Antiviral agents

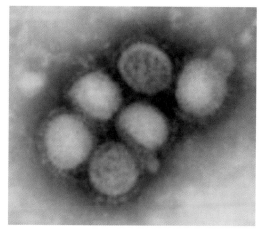

Influenza virus

KEY FACTS

- A significant number of important viral diseases are controlled by vaccination but for some it has not been possible to develop effective vaccines.
- Antiviral agents target various stages in the viral replication cycle but as this is intimately linked to the functioning of the human cell their use is often associated with toxicity.
- Because of the potential for toxicity, systemic antiviral agents are only used for severe infections.
- Many antiviral agents act by interfering with DNA synthesis and of these an important group are the nucleoside analogues.
- For infections such as herpes and HIV it is not possible to completely eliminate the virus from the body; the antiviral agents merely suppress replication.

12.1 Viral infections

Bacteria are prokaryotic cells and so possess several unique features that can act as chemotherapeutic targets. These include the peptidoglycan in the cell wall and 70s ribosomes, which are not found in human cells. Many of the currently available antibacterial antibiotics target these sites and so the side effects which result from their use tend to be minimal. Viruses, on the other hand, rely on the metabolic machinery of human cells for their replication and so if we are to interfere with this process using antiviral agents it is almost inevitable that adverse effects will occur.

Table 12.1 lists the most common viral infections encountered in developed countries. Vaccines are available for many of these viruses and have provided the most effective means for the control of their associated infections. However, vaccines for the control of some viruses have proved elusive, particularly those such as rhinoviruses and adenoviruses, which give rise to the common cold. These infections are usually mild and self-limiting and so do not warrant intervention with potentially toxic antiviral agents. Vaccines are not currently available for the herpes group of viruses or the HIV responsible for AIDS, but here the disease can be so severe, possibly even life-threatening, that the risk associated with antiviral chemotherapeutic agents is justified. Other infections such as influenza, hepatitis and respiratory syncytial disease are also sufficiently serious to warrant treatment using antiviral agents.

Essential Microbiology for Pharmacy and Pharmaceutical Science, First Edition. Geoffrey Hanlon and Norman Hodges.
© 2013 John Wiley & Sons, Ltd. Published 2013 by John Wiley & Sons, Ltd.

Table 12.1 Commonly encountered viral infections and the methods available to treat them.

Disease	Vaccine availability	Antiviral chemotherapy
Polio	+	−
Measles	+	−
Rubella	+	−
Mumps	+	−
Influenza	+	+
Hepatitis	+	+
Rhinovirus, adenovirus, Coxsackie virus, etc.	−	−
Respiratory syncytial virus	−	+
Herpes simplex virus types I and II	−	+
Cytomegalovirus	−	+
Epstein–Barr virus	−	+
Varicella zoster virus	+	+
Human immunodeficiency virus	−	+

12.2 Targets for antiviral agents

Figure 12.1 shows a diagram of the infection of a cell by the human immunodeficiency virus, which gives rise to AIDS. In general the process of viral replication can be summarized as follows:

1. Adsorption of virus to host cell and entry.
2. Uncoating to liberate viral genome.
3. Synthesis and/or replication of viral DNA.
4. Integration of DNA into host genome (for latent viruses).
5. Production and assembly of new viral components (nucleic acid and protein).
6. Maturation.
7. Release of new virions.

Each of these stages represents a potential intervention site for antiviral therapy, and their usefulness will be discussed in this chapter. However, to date, most of the therapeutic strategies have been directed towards interference with the replication of viral DNA and this will be dealt with first.

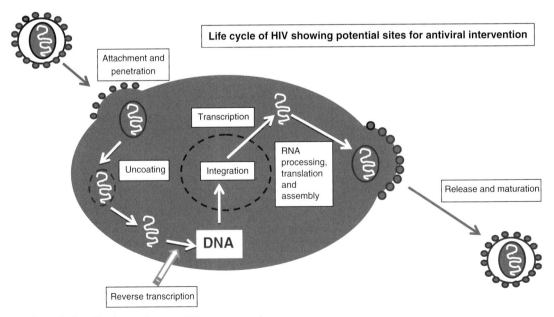

Figure 12.1 Viral replication cycle using HIV as an example.

Table 12.2 Different members of the herpes group of viruses.

Virus	Disease
Herpes simplex type I	Cold sores
Herpes simplex type II	Genital herpes
Varicella zoster / Herpes zoster	Chicken pox leading on to shingles
Cytomegalovirus	Asymptomatic in normal people. Retinitis, pneumonia and colitis in immunosuppressed patients. Congenital infection in foetus if contracted during pregnancy. Can lead to foetal damage.
Epstein–Barr virus	Glandular fever

and the analogue is idoxuridine. We can see that the pentose sugar is unchanged but the purine base has an iodine atom instead of a methyl group in the idoxuridine molecule. When this is administered it is taken up by the DNA polymerase and used to make DNA. However, the idoxuridine cannot base-pair properly and so the function of the resulting DNA is compromised.

Idoxuridine was one of the first nucleoside analogues and was synthesized in 1959 as part of an intensive search for anticancer drugs. Its anticancer activity was quite weak but it demonstrated potent anti-viral activity. It replaces thymidine in the growing DNA chain and can still form DNA chains as it possesses −OH groups at the 5 and 3 positions on the pentose sugar. Important points to note about idoxuridine are as follows:

- its main use is in the treatment of superficial herpes simplex infections of the skin and eye;
- it is too toxic to be used systemically;

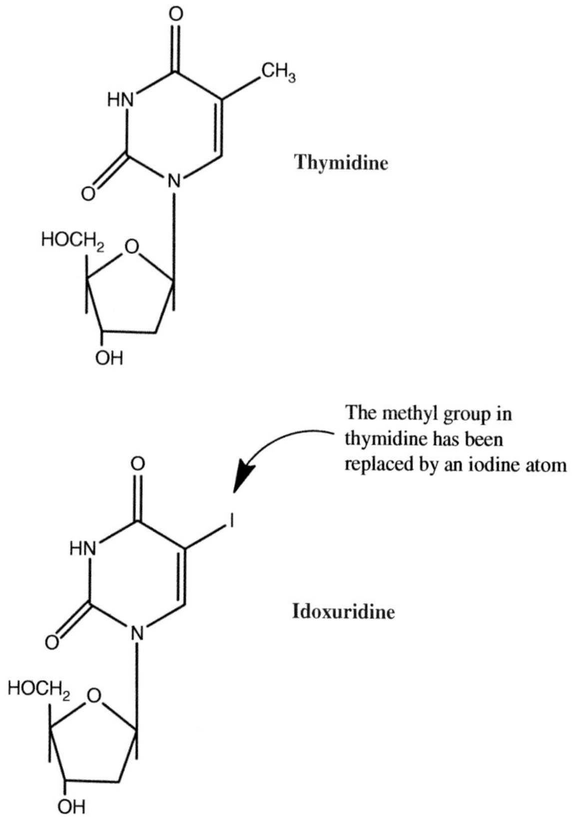

12.3 Antiviral agents active against the herpes group of viruses

Modern antiviral agents were developed initially for the treatment of infections caused by the herpes group of viruses, in particular herpes simplex type I, and so these will be considered first. The herpes group contains a range of latent viruses as can be seen in Table 12.2 . These viruses give rise to diseases which range in severity from simply being a nuisance to being life-threatening.

12.3.1 Nucleoside analogues

Antiviral therapy is aimed at preventing viral DNA from being replicated within the host cell. The strategy is to use molecules which resemble natural nucleosides but which are modified to make them nonfunctional. These are called *nucleoside analogues*.

> If you would like to remind yourself about the structure of DNA and the functioning of these agents then you should refer to the RECAP section at the end of this chapter.

An example of a nucleoside analogue is shown in Figure 12.2 . Here, the natural nucleoside is thymidine

Figure 12.2 The close relationship between the nucleoside analogue idoxuridine and the natural compound thymidine.

- it causes adverse effects to kidney, bone marrow and liver;
- it cannot kill latent viruses – it only prevents replication;
- it has now been superseded by other less toxic drugs.

In the search for less toxic nucleoside analogues, aciclovir was developed. Aciclovir is a substituted guanine derivative that lacks an −OH group at position 3 on the sugar so it cannot accommodate chain elongation. Aciclovir itself is inactive and it must be phosphorylated to form the nucleoside triphosphate in order to be utilized by DNA polymerase.

In virus-infected cells aciclovir is phosphorylated by the viral enzyme thymidine kinase to form aciclovir monophosphate. Subsequent phosphorylation to di- and tri- phosphates is carried out by cellular kinases. In noninfected cells thymidine kinase does not exist and

so aciclovir is not phosphorylated and remains inactive. See Figure 12.3.

Points to note about the uses of aciclovir are as follows:

- It is used in the treatment of herpes infections of the skin and mucous membranes:
  - herpes simplex virus types I and II;
  - herpes simplex keratitis;
  - varicella zoster (chicken pox and shingles).
- It can be given orally, topically, or by intravenous infusion.
- Its solubility is poor, hence it should not be given by bolus injection – it is infused slowly over 1 hour.
- The pH of the solution is high (11) and may lead to irritation at the site of infusion.
- Poor solubility leads to poor absorption by mouth so high doses are needed:

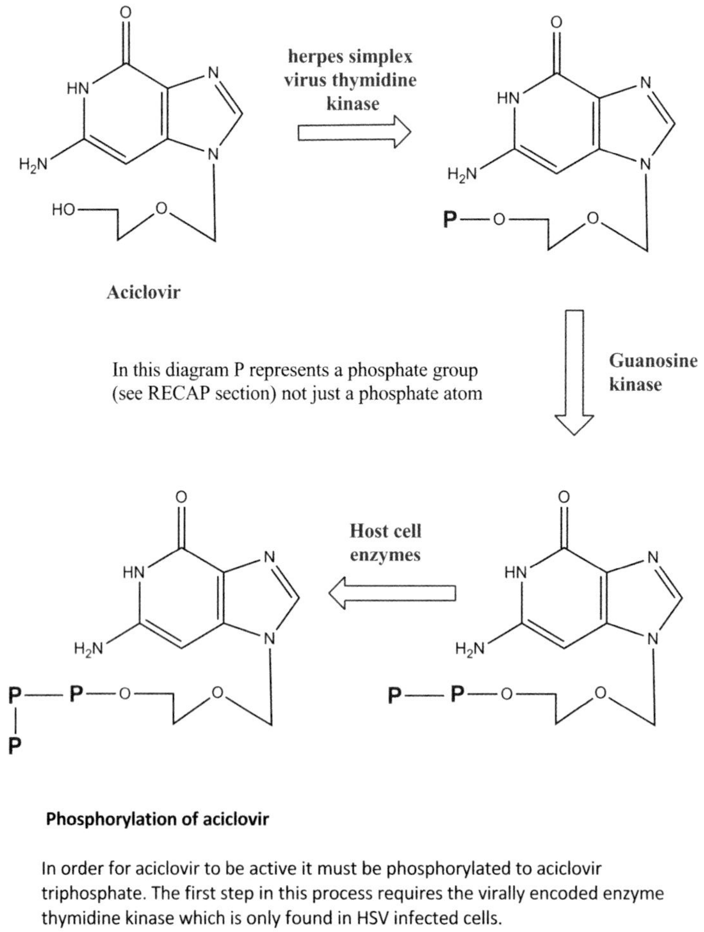

Phosphorylation of aciclovir

In order for aciclovir to be active it must be phosphorylated to aciclovir triphosphate. The first step in this process requires the virally encoded enzyme thymidine kinase which is only found in HSV infected cells.

Figure 12.3 Intracellular phosphorylation of aciclovir.

- 200 mg–400 mg five times per day for HSV type I.
- 800 mg five times per day for varicella and herpes zoster.

- Clinically, it is highly effective with minor side effects.
- Newer drugs such as famciclovir and valaciclovir are similar to aciclovir and differ mainly in their improved oral absorption.

A major problem with aciclovir is its relative lack of activity against a formidable member of the herpes group called cytomegalovirus. This does not cause many problems in otherwise healthy people but can cause blindness or even be life threatening in immunocompromised patients.

Ganciclovir is very similar in structure to aciclovir and its mode of action is the same. However, it is a much more efficient substrate than aciclovir for viral thymidine kinase and is converted to di- and tri- phosphates much more rapidly (10 × more ganciclovir triphosphate than aciclovir triphosphate in equivalently treated cells). Ganciclovir is broken down and eliminated more slowly from infected cells and it does not lead to chain termination. As a consequence of these factors, ganciclovir is much more toxic than aciclovir and adverse reactions have been reported in nearly half of all patients as follows:

- leucopenia (abnormal diminution of white blood cells) reported in 25% of AIDS patients and 15% of non-AIDS patients;
- thrombocytopenia (abnormal diminution of platelets) reported in 6% of AIDS patients and 12% of non-AIDS patients;
- numerous other less severe reactions including rash, problems at infusion sites and CNS problems;
- danger of teratogenicity (foetal damage), impairment of fertility, mutagenicity and potential carcinogenicity.

As a result ganciclovir is recommended only for the treatment of life-threatening or sight-threatening cytomegalovirus infections.

12.4 Inorganic pyrophosphate mimics

The RECAP section at the end of this chapter explains that, when DNA polymerase adds a nucleoside triphosphate to a growing DNA chain, two of the phosphates are cleaved (inorganic pyrophosphate) to provide energy. From this reaction we can see that if the inorganic pyrophosphate was present in excess then the action of the DNA polymerase would be inhibited. The antiviral agent foscavir has structural similarities to pyrophosphate and inhibits the enzyme by mimicking the end product of the reaction. Its interaction is fairly specific for viral DNA polymerases but it does interact with cellular DNA polymerases at higher concentrations. Consequently, it is a very toxic drug and renal impairment can occur in up to 50% of patients. It is therefore only indicated for the treatment of CMV retinitis in patients with AIDS and in whom ganciclovir is contraindicated or inappropriate.

12.5 Anti-HIV drugs

Acquired immune deficiency syndrome (AIDS) was first recognized in 1981 and is caused by the human immunodeficiency virus (HIV), a retrovirus which attacks T lymphocytes and other cells of the immune system (Chapter 8). The HIV has a slightly different infection cycle from the viruses illustrated in Chapter 6 (see Figure 12.1). It is called a retrovirus because its genome is in the form of RNA, and on entering the host cell this RNA is converted to DNA by the viral enzyme reverse transcriptase, which is the equivalent enzyme to DNA polymerase. The DNA is then integrated into the host cell genome where it resides permanently as a latent virus.

The HIV has surface proteins which act as binding sites; these are called gp120 and gp41. In addition, there are receptor sites on the lymphocyte surface. The main receptor on the lymphocyte is CD4, to which the gp120 protein attaches. Other receptors are CXCR4 or CCR5, to which the gp41 protein binds. An HIV infection of lymphocytes requires attachment at both sites and tight attachment of the virus to the host surface receptors leads to membrane fusion. The HIV RNA contains nine genes, which code for structural proteins including capsid proteins but also three key enzymes:

- reverse transcriptase;
- integrase;
- protease.

These enzymes have provided useful targets for antiviral chemotherapy.

12.5.1 Nucleoside reverse transcriptase inhibitors

Those nucleoside analogues acting at this target site are referred to as nucleoside reverse transcriptase inhibitors (NRTIs). Zidovudine otherwise known as 3-azido-3-

deoxythymidine (AZT) was the first of these to be developed. It is a thymidine analogue where the 3-OH group is replaced by an azido group (N3). Unlike aciclovir, AZT is converted to the triphosphate form by cellular enzymes in both noninfected and infected cells where it is utilized by reverse transcriptase and acts as a chain terminator because the azido group cannot form a link with the phosphate group on an adjacent nucleotide. Cellular DNA polymerase is approximately a hundred-fold less susceptible than reverse transcriptase and hence side effects are minimized. Zidovudine is well absorbed from the gut and so can be taken orally. However, it does have profound toxic effects, particularly causing blood disorders, so patients should be assessed for haematological toxicity and their dosage adjusted accordingly.

An important point to note about all these anti-HIV drugs is that they do not kill the virus; they simply prevent its replication. Once a patient has become infected with a latent virus there is at present no therapy that will eradicate it.

In an attempt to find less toxic NRTIs, didanosine and zalcitabine were developed in the 1990s.

Facts about didanosine:

- it is acid labile often given on an empty stomach or after antacids;
- patients show raised CD4 cell counts and decreased HIV-1 RNA counts;
- its spectrum of adverse reactions differs from AZT:
  – haematological toxicity is minimal;
  – there are problems with peripheral neuropathy (damage to nerves) and pancreatitis.

Facts about zalcitabine:

- its absorption rate is decreased by the presence of food;
- improvements are shown in surrogate markers;
- significant clinical improvements arise in those who can tolerate the drug;
- haematological toxicity is rare, but peripheral neuropathy common;
- pancreatitis may occur and is serious but rare.

Lamivudine is a more recently developed NRTI, which has a number of significant advantages:

- it is a potent antiviral agent;
- it is rapidly absorbed – bioavailability > 80%;
- it causes much less peripheral neuropathy than previous drugs;
- it is clinically well tolerated drug with the most common side effects being headache, fatigue and diarrhoea.

12.5.2 Nucleotide reverse transcriptase inhibitors

These are *nucleotide* analogues (as distinct from nucleoside analogues), which means that they are already phosphorylated. Other than that they work in the same way as NRTIs and include the antiviral agent tenofovir.

12.5.3 Non-nucleoside reverse transcriptase inhibitors (NNRTIs)

Another group of drugs acting on reverse transcriptase are the non-nucleoside reverse transcriptase inhibitors. As the name suggests, these are not nucleoside analogues and their structures are completely different. They bind at a different site on the enzyme to nucleoside analogues and there is no equivalent receptor on human enzymes. Examples include nevirapine and efavirenz.

12.5.4 Protease inhibitors (PIs)

It was noted above that HIV produces a number of enzymes – one of which is protease. This is responsible for modifying the newly formed proteins produced during viral replication so that they are in the correct configuration for incorporation into the developing viral progeny. This was recognized as a potential target, and protease inhibitors were first introduced in 1996. Unlike nucleoside analogues, they do not require phosphorylation for activity. They prevent post translational modification of viral proteins leading to production of noninfectious virus particles. Fortunately, these inhibitors do not interfere with activity of human proteases, for example trypsin and pepsin.

When used in combination with other agents they bring about a dramatic improvement in the clinical picture and reduction in HIV-related deaths. There are problems with lipodystrophy – abnormal distribution of body fat – and they can also give elevated plasma lipid levels. Caution is needed in patients with haemophilia who are at increased risk of bleeding. There are now numerous examples in clinical use including ritonavir, saquinavir and lopinavir.

12.5.5 Fusion inhibitors

The HIV is an enveloped virus and has to fuse membranes with the host cell in order to bring about an infection. We have already said that two receptors on

HIV (gp120 and gp41) bind to host cell receptors (CD4 and CXCR4) respectively and are responsible for infection. If these receptors could be blocked then the infection process will be stalled.

Enfuvirtide is a 36 amino acid peptide derived from HIV gp41 and inhibits gp41-mediated fusion.

It works extracellularly and has to be self-administered by subcutaneous injection. The most common side effect is injection-site reaction. It is synergistic with NRTIs, NNRTIs, PIs and some other antiretrovirals. There appears to be no cross-resistance with other antiretroviral classes and the drug is active against virus resistant to other classes of antiretrovirals. Resistance can occur in the binding area of gp41.

Enfuvirtide is not used routinely but is licensed for use in infections that have not responded to the regimen of other antiretrovirals.

Maraviroc is an orally active fusion inhibitor that has recently been released onto the market. It is only active against viruses binding to the CCR5 co-receptor; not CXCR4. The reason for the difference in virus receptor binding is unclear but CXCR4 binding variants may emerge as the disease progresses. Patients on maraviroc have been shown to be twice as likely to achieve undetectable HIV-1 RNA levels and double the gain in their CD4+ T cells. The profile of adverse effects is also acceptable.

12.5.6 Integrase inhibitors

After insertion of HIV viral RNA into the host cell, the genetic material is converted to dsDNA by the enzyme reverse transcriptase produced by the virus. This DNA is then further processed by another viral enzyme called integrase which produces sticky ends on the DNA of virus

and host. The viral DNA is then spliced into the host DNA (see Figure 12.4).

Raltegravir is an integrase inhibitor that has recently been released for clinical use.

12.5.7 Maturation inhibitors

Prior to release from the host cell all newly constructed HIV particles undergo a maturation process and this is a key step in viral replication. Development is underway on a group of drugs called maturation inhibitors which disrupt this late stage viral maturation process. One of the potential targets is the HIV Gag protein, which forms the capsid shell of the virus. Maturation inhibitors cause Gag protein, and hence the capsid, to be defective and noninfectious. These drugs have been shown to be active against drug-resistant strains of HIV because they act at a different site but as yet none has been licensed for use.

12.6 HIV treatment: Highly Active Antiretroviral Therapy – HAART (British HIV Association Guidelines)

The British HIV Association gives recommendations on the most appropriate treatment regimens for HIV. The HAART regimens need to be individualized to achieve maximum potency, durability, adherence and tolerability and the goal is to achieve a viral load of < 50 copies per ml within 4–6 months of starting treatment. The current recommendations are shown in Table 12.3 and as can be seen a number of options are available depending upon

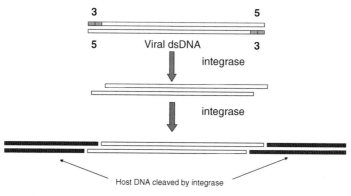

Figure 12.4 The virally encoded integrase enzyme is responsible for splicing the newly produced HIV DNA into the host cell genome.

Table 12.3 British HIV Association Guidelines on the management of HIV.

Regimen	A	B	C
Preferred	Efavirenz	Tenofovir	Lamivudine
		Abacavir	Emtricitabine
Alternative	Lopinavir/r	Didanosine	
	Fosamprenavir/r	Zidovudine	
	Atazanavir/r		
	Saquinavir/r		
Specific groups	Nevirapine[a]		
	Atazanavir[b]		

Choose one drug from columns A, B and C.
[a]Only when CD4 <250cells/ml in females and <400cells/ml in males.
[b]Where there are established cardiovascular risk factors or a PI is required.

individual patient circumstances. There is a need to monitor viral load reduction closely in the early weeks of treatment.

12.7 Viral hepatitis

Hepatitis is inflammation of the liver primarily caused by hepatitis A, B or C viruses. Hepatitis A mainly occurs in epidemics among children and young adults in institutions. Hepatitis B virus (HBV) infection occurs by direct inoculation into blood or contamination of mucous membranes and hepatitis C resembles a mild form of hepatitis B, commonly associated with transfusion of blood or blood products.

Hepatitis B may be self-limiting but chronic liver disease can develop in 10% of patients and chronic infection is associated with an increased risk of hepatic carcinoma.

Facts about hepatitis B:

- It is estimated that there are 300 million chronic carriers worldwide.
- Carriers may be asymptomatic or have chronic active hepatitis, cirrhosis and/or carcinoma.
- Immunization is available against hepatitis A and B infections.
- Antiviral chemotherapy for hepatitis B includes:
  - Interferon α – limited usefulness (only 50% response rate).
  - Lamivudine – used for initial treatment; resistance may occur on long-term treatment.
  - Entecavir – nucleoside analogue. Not active against HIV, but stops HBV from replicating in the liver.

- Tenofovir – a nucleotide analogue used as first-line therapy.
- Adefovir dipivoxil – a nucleotide analogue effective in lamivudine-resistant chronic hepatitis B.

12.8 Influenza

Influenza is a respiratory illness caused by an influenza virus and the symptoms include headache, fever, cough, sore throat, aching muscles and joints. There are three main types: influenza A, B and C, of which influenza A is the most important. Infections occur mostly in the winter and peak around December to March in the northern hemisphere. Illnesses occurring in summer are usually due to other viruses. For most patients influenza is an unpleasant illness but they will recover fully; for others it can be life threatening. Common complications are bronchitis and secondary bacterial pneumonia, which are a particular problem for the elderly, asthmatics and those in poor health.

Table 12.1 shows that influenza vaccination is available and this forms the primary approach in the control of this disease in the elderly and at-risk population. However, antiviral agents are also used to manage epidemic outbreaks in nonimmunized patients.

Influenza viruses A and B are antigenically unstable. Their viral surface contains haemagglutinins and neuraminidase, which constantly mutate. The population develops immunity to the viruses currently in circulation but occasionally a major antigenic change occurs and epidemics or even pandemics result. The World Health Organisation (WHO) monitors the antigenic makeup of influenza viruses and recommends the formulation of vaccines to the manufacturers.

Oseltamivir and zanamivir inhibit the neuraminidase enzyme on the influenza viral envelope.

Neuraminidase is essential for the release of virus from its host and the enzyme hydrolyses the sialic acid end of glycoproteins and glycolipids present on the surface of host-cell membranes. Oseltamivir is a sialic acid analogue mimicking the substrate and binding to the active site on neuraminidase.

Oseltamivir (Tamiflu) is orally active while Zanamivir (Relenza) is available only for administration by inhalation. Both are licensed for treatment if started within 48 hours and will reduce the duration of symptoms by between 1 and 1.5 days.

12.9 Respiratory syncytial virus

Respiratory syncytial virus (RSV) is the most frequent cause of serious respiratory disease in infants and young children. It occurs as bronchiolitis, pneumonia or upper respiratory-tract infections and can also cause severe disease in adults, especially the elderly and immunocompromised. Patients with underlying cardiac or pulmonary disease are also at risk. The disease is usually self-limiting (10–14 days) but may require hospitalization.

Respiratory syncytial virus is a paramyxovirus with a ssRNA genome enclosed in a protein envelope. There are two surface glycoproteins, G and F. G glycoprotein mediates attachment to host cell receptors while F glycoprotein induces viral penetration and fusion. F glycoprotein also induces the fusion of viral-infected cells, resulting in *syncytium* formation, which are multi-nucleated giant cells, giving the virus its name.

Outbreaks occur in the autumn, winter and early spring and nosocomial outbreaks in paediatric wards and nurseries are a major problem. The virus replicates in the ciliated epithelial cells of the middle and lower respiratory tract and disrupts protein and nucleic acid synthesis leading to cell death. Enzymes are then released which activate complement and initiate local inflammatory responses. Problems are caused when necrotic cells, mucus and fibrin plug the small airways.

Antibodies are formed to G and F glycoproteins but immunity is incomplete and repeated episodes of infection are common.

12.9.1 Ribavirin

Ribavirin is a nucleoside analogue where the base is neither purine nor pyrimidine. It is effective against a range of DNA and RNA viruses and is licensed for administration by inhalation for the treatment of severe bronchiolitis caused by RSV in infants, especially when they have other serious diseases. However, there is no evidence that ribavirin produces clinically relevant benefit in RSV bronchiolitis.

12.9.2 Palivizumab

Palivizumab is a humanized monoclonal antibody composed of human (95%) and murine (5%) antibody sequences. It was the first clinically used human antibody approved by the FDA in 2002.

Activity of the antibody is directed against the F protein on the outer surface of the RSV. This reduces viral activity and cell-to-cell transmission of the virus, and blocks the fusion of infected cells.

By binding to this target on RSV, palivizumab prevents viral invasion of the host cells in the airway.

It is indicated for the prevention of lower respiratory tract diseases caused by the RSV in children usually younger than 2 and it should also be considered for children less than 6 months old with congenital heart disease or with pulmonary hypertension.

Table 12.4 summarizes the different viral infections and illustrates the antiviral agents available to manage them.

Table 12.4 Summary of antiviral chemotherapy.

Disease	Class	Examples	Mode of action
Herpes simplex	Nucleoside analogues	Idoxuridine Aciclovir Famciclovir Valaciclovir Ganciclovir	Interfere with DNA polymerase
	Inorganic pyrophosphate mimics	Foscavir	
HIV/AIDS	Nucleoside reverse transcriptase inhibitors (RTIs)	Zidovudine Zalcitabine Didanosine Lamivudine Emtricitabine Abacavir Stavudine	Nucleoside analogues that interfere with reverse transcriptase (RT)
	Nucleotide reverse transcriptase inhibitors	Tenofovir	Similar to nucleoside analogue RTIs but do not need to be phosphorylated
	Non-nucleoside reverse transcriptase inhibitors	Nevirapine Efavirenz Etravirine	Interfere with RT by different mechanism from above
	Protease inhibitors	Amprenavir Darunavir Fosamprenavir Indinavir Nelfinovir Lopinavir Ritonavir Saquinavir Tipranavir	Prevent post translational modification of viral proteins
	Fusion inhibitors	Enfuvirtide Maraviroc	Bind to host cell and block viral attachment
	Integrase inhibitors	Raltegravir	Block integration of viral DNA into host genome.
	Maturation inhibitors	Still under development	Disrupt late stage viral maturation.
Influenza	Neuraminidase inhibitors	Oseltamivir Zanamivir	Prevent viral release by inhibiting neuraminidase enzyme on virus.
Hepatitis B	Cytokines	Interferons	
	Nucleoside analogues	Lamivudine	See above
	Nucleotide analogues	Adefovir dipivoxil	See above
Respiratory syncytial virus	Nucleoside analogue	Ribavirin	See above
	Humanized monoclonal antibody	Palivizumab	Binds to F glycoprotein on surface of RSV

Acknowledgement

Chapter title image: PHIL ID #11212; Photo Credit: C.S. Goldsmith and A. Balish, Centers for Disease Control and Prevention.

RECAP section explaining the structure of nucleic acids and the process of DNA replication

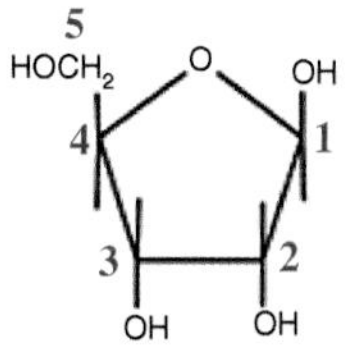

D-ribose 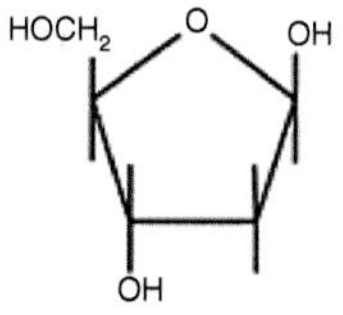   2-deoxy D-ribose

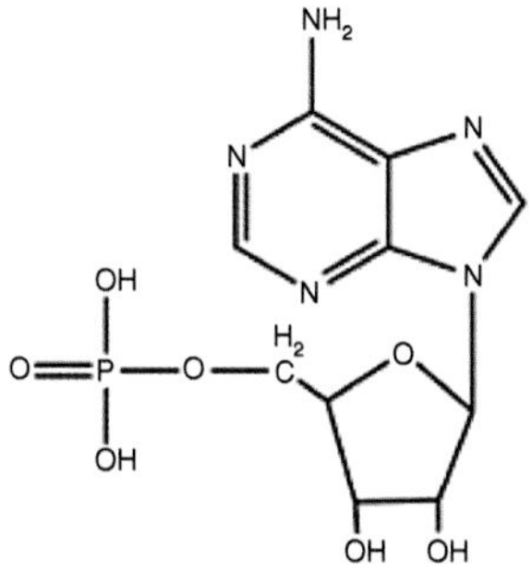

Adenosine-5-monophosphate

Nucleoside = purine or pyrimidine base linked to a pentose sugar
Nucleotide = base + sugar + phosphate group

RECAP section

In order to understand how these antiviral agents work, we need to remind ourselves of the structure of DNA and the process of DNA replication.

DNA is composed of nucleotides, which comprise a sugar, an organic base (purines or pyrimidines) and a phosphate group (see figure above). The numbering of the carbon atoms in the pentose sugar is important and we can see that the organic base is attached to carbon atom number 1 and the phosphate group is attached to the carbon atom in position 5. The carbon atom at position 3 is used to join adjacent nucleotides together to form a chain. In this process the phosphate group of one nucleotide links to the 3-OH group on another and so a chain of nucleotides can be built up with a sugar/phosphate backbone. The organic bases are seen to stick out sideways. In fact, single chains of DNA do not usually occur and another chain normally binds along the length of the molecule to form a double strand where the organic bases hydrogen bond in a specific manner. Adenine always pairs with thymine and cytosine always pairs with guanine. Note that the individual chains have a direction as a result of the orientation of the pentose sugar. One of the chains will be in the 3–5 direction while the partner chain lies in the 5–3 direction.

The process of DNA replication is rather like placing bricks in a wall. DNA polymerase is the enzyme which drives the construction of new DNA molecules using the existing DNA strand as a template and adding nucleotides in a stepwise manner onto a free 3-OH group on the end nucleotide of the growing chain. The nucleotides which are added by the DNA polymerase are in the triphosphate form and, during the addition process, inorganic pyrophosphate is released providing energy to drive the reaction forwards.

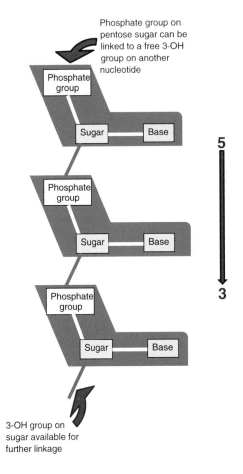

Phosphate group on pentose sugar can be linked to a free 3-OH group on another nucleotide

Phosphate group

Sugar — Base

5

Phosphate group

Sugar — Base

Phosphate group

Sugar — Base

3

3-OH group on sugar available for further linkage

NUCLEOSIDE

Purine or pyrimidine base linked to a pentose sugar

NUCLEOTIDE

Purine or pyrimidine base linked to a pentose sugar containing phosphate groups.

Nucleotides join together to form a chain with a sugar/phosphate backbone.

The chain has direction due to the orientation of the 5 and 3 groups on the pentose sugars.

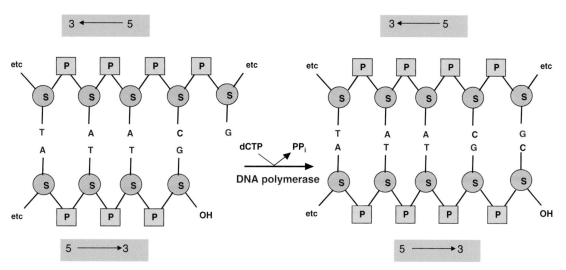

DNA replication - role of DNA polymerase

The DNA polymerase enzyme acts like a brick-layer just adding bricks to the DNA wall. In the example above it takes a nucleotide triphosphate containing the appropriate base (in this case cytosine) and combines it with the free 3-OH group present on the last pentose sugar. The cytosine base pairs with the corresponding guanine base and inorganic pyrophosphate is released. A further 3-OH group is generated, to which another nucleotide can be added.

The orientation of the nucleotide chains (shown in grey) refers to the relative positions of the carbon atoms on the pentose sugars.

Chapter 13
Antibiotic resistance

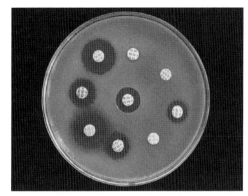

Measuring antibiotic sensitivity

KEY FACTS

- Many bacteria are naturally resistant to certain antibiotics, either because they do not possess the cell target with which the antibiotic interferes or because the antibiotic cannot enter the cell.
- Of more importance is the situation in which bacteria *become* resistant to an antibiotic as a result of a mutation in the genes they already possess or the acquisition of new resistance genes from another cell.
- Bacterial conjugation in which genetic information is transferred from one cell (a donor) to another (recipient) is the most important mechanism of gene transfer.
- Antibiotic sensitivity is usually measured using antibiotic-impregnated paper discs that create zones of growth inhibition on agar in Petri dishes inoculated with test bacteria.
- Antibiotic resistance should be regarded as the rule, not an exceptional occurrence. Most antibiotics become progressively less effective over time.
- Bacteria achieve resistance by several mechanisms: degrading the antibiotic itself; modifying the cell permeability properties so the antibiotic cannot enter the cell; modifying the cellular target so the antibiotic cannot bind to it; or by efflux pumps which actively expel antibiotic molecules from the cell.
- The production of β-lactamases and, particularly, the extended-spectrum β-lactamases is a particularly important and problematic resistance mechanism.

Resistance describes the situation in which an antibiotic fails to kill or inhibit the growth of a microorganism at concentrations that can safely be achieved at the site of infection in the body. From the attention that the subject receives in the media it might seem that antibiotic resistance is a new phenomenon, but that is not the case; penicillinase, the enzyme that enables many bacteria to resist the action of β-lactam antibiotics, was first described in 1940 – before penicillin came into clinical use. It follows from this that the bacteria already possessed the ability to produce penicillinase even before penicillin was discovered: despite their rapid rate of reproduction, the ability to produce the enzyme could not have evolved in the bacteria in response to penicillin exposure since its discovery in 1928. Evolution occurs over a much longer time period, so it is probable that resistance genes evolved over millennia, particularly amongst soil-dwelling bacteria, as a means of protection against antibiotics produced by other organisms that are also present in the soil, e.g. *Penicillium*, *Streptomyces* and *Bacillus* species. There are, therefore, bacteria that are naturally resistant to many, or even most, of the commonly used antibiotics and, for all practical purposes, it is true to say that they have always been resistant, and are

Essential Microbiology for Pharmacy and Pharmaceutical Science, First Edition. Geoffrey Hanlon and Norman Hodges.
© 2013 John Wiley & Sons, Ltd. Published 2013 by John Wiley & Sons, Ltd.

always likely to be in the future; this is described as *innate* or *intrinsic* resistance.

What is much more of a problem, and the reason for the media attention, is the fact that so many organisms that were originally sensitive to particular antibiotics when the drugs were first discovered are sensitive no longer, so the antibiotics are becoming less useful. This is described as *acquired* resistance because it originates from the organisms acquiring new genes, either by mutations of those they already possess or, more problematically, from other microorganisms. This transmission of genes from one cell to another without reproduction or increase in cell numbers is termed horizontal transmission, whereas the term vertical transmission describes genes simply being passed through the generations from each cell to its offspring.

Media coverage of antibiotic resistance might also create the impression that it is a relatively infrequent occurrence, but that is not so either; resistance is the rule, not the exception. As with penicillin, the first reported cases of resistance to new antibiotics usually occur before they are in widespread use – often during clinical trials – and there are few antibiotics that have remained as potent during their lifetime as they were initially. Usually the concentrations required to inhibit the growth of the target organisms rise slowly over the course of several years as a result of the cumulative effects of minor increases in resistance arising from mutations; this phenomenon is known as 'resistance creep' and it illustrates another aspect of resistance: it is not an all-or-none phenomenon whereby an organism changes overnight from being totally sensitive to being totally resistant. There are intermediate states; so, for example, *Staphylococcus aureus* can be categorized as vancomycin resistant, vancomycin sensitive and vancomycin intermediate (which might only be expected to respond to treatment with higher than normal doses).

13.1 Innate (intrinsic) resistance

For an antibiotic to kill or inhibit the growth of a bacterial cell, two conditions must be satisfied: the cell must possess the antibiotic 'target' and the antibiotic must reach the target in sufficiently high concentration to inhibit its operation. Targets are usually, but not invariably, enzyme systems. Common examples include:

- enzymes responsible for peptidoglycan synthesis (termed penicillin binding proteins – PBPs – which are inhibited by all β-lactam antibiotics);

- those synthesizing or reducing folic acid (inhibited by sulphonamides and trimethoprim);
- and the DNA gyrase enzymes with which fluoroquinolones interact.

If an organism does not possess the target it will be innately resistant; for example, mycoplasmas do not synthesize peptidoglycan in their cell walls so they are resistant to penicillins. If the antibiotic cannot reach the potential target because of permeability barriers, that, too, confers innate resistance, for example the vancomycin molecule is simply too large to enter the cells of most Gram-negative bacteria so its action is mainly limited to Gram-positive species.

13.2 Measurement of resistance

Antibiotic resistance (or susceptibility) is most commonly measured using antibiotic-impregnated paper disks that are placed on the surface of inoculated Petri dishes. During incubation the antibiotic dissolves in the gel and diffuses outwards from the disk to give a concentration gradient that produces a zone of growth inhibition around the disk (Figures 13.1A and 16.2). Susceptibility-testing disks impregnated with all of the commonly used antibiotics are commercially available from several manufacturers, and the amount of antibiotic on the disk is adjusted so that the concentration range within the agar is similar to that in body fluids after the administration of standard therapeutic doses.

The size of the inhibition zone is measured and compared with published tables (see, for example, the UK Society for Antimicrobial Chemotherapy web pages (http://bsac.org.uk/susceptibility/guidelines-standardized-disc-susceptibility-testing-method), which indicate whether the organism in question is sensitive or resistant. The zone diameter is influenced by several factors including the pH and composition of the culture medium, the inoculum concentration and the incubation conditions, and although there is no internationally accepted method for conducting antibiotic susceptibility tests there are several well documented variants of the basic procedure that standardize these factors. Instruments are available both for the semiautomatic preparation of the test plates and for reading zone diameters. In some cases a reference strain is inoculated onto one half of the plate and the test strain onto the other half, so that the effects of variations in experimental conditions can be discounted. In Figure 13.1B, for example, the *Enterobacter*

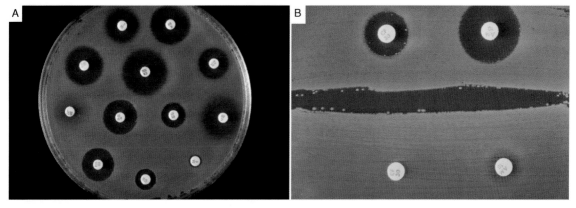

Figure 13.1   Antibiograms showing: A. differing susceptibility of an *Enterobacter* species to various antibiotics; B. a close-up of a test plate showing the sensitivity of two *Enterobacter* species to cephalothin (left disks) and ampicillin (right) – see text. *Source:* PHIL ID #3031 and #3032; Photo Credit: Dr. J.J. Farmer, Centers for Disease Control and Prevention.

species on the top half of the plate is clearly sensitive to the two antibiotics cephalothin and ampicillin, whereas that on the bottom is resistant.

Occasionally antibiotics cannot conveniently be tested in this way because they require unusual incubation conditions or they do not diffuse well in agar gels (which contain 97–98% water). Some antifungal agents, for example, are poorly water soluble and are better tested by minimum inhibitory concentration (MIC) tests, which, as their name suggests, determine the lowest concentration of the antibiotic that is effective in inhibiting growth of the test organism. Again, there is no standardized method, so the test can be performed in a variety of ways, but usually the antibiotic is incorporated in tubes or other containers of liquid medium at a range of doubling dilutions (for example 16, 8, 4, 2, 1, 0.5 etc. mg/L) and all of the tubes inoculated with the test organism. The MIC is the lowest concentration at which there is no growth after incubation. Figure 13.2 shows an MIC determination using a microtitre plate in which the antibiotic concentration decreases from left to right (excluding columns 11 and 12 on the right which are controls); in row E, for example, there is no growth of the test organism in wells 1–4, but growth thereafter in wells 5–10 (shown by cloudiness of the liquid), so the MIC is the concentration in well E4. If the antibiotic concentration that can safely be achieved at the infection site does not exceed the measured MIC the organism is regarded as resistant and so-called antibiotic 'breakpoints' are used for the prediction of successful therapy. A breakpoint is an MIC threshold, and organisms having an MIC below this threshold value can be expected to be inhibited or killed by standard doses of the antibiotic.

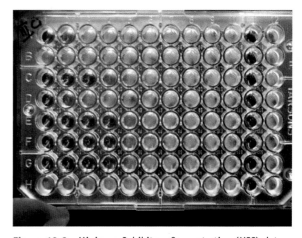

Figure 13.2   Minimum Inhibitory Concentration (MIC) determination of an isolate by microbroth dilution technique in a 96-well microtitre plate. *Source:* http://commons.wikimedia.org/wiki/File:MIC_microbroth_dilution.jpg.

13.3 Origins of antibiotic resistance

In addition to 'innate' and 'acquired', there are other terms used to describe the characteristics of antibiotic resistance, and a distinction is often made between resistance of phenotypic and genotypic origin (Figure 13.3). A phenotypic change is, by definition, one that does not arise from an alteration in the genes the organism possesses; rather, it is one in which the cells in

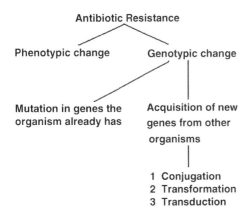

Figure 13.3   Origins of antibiotic resistance.

a population, such as pathogenic bacteria at an infection site, modify their physical structure or biochemical properties in response to an environmental stress, for example exposure to antibiotic. This is sometimes referred to as 'adaptive resistance', and it is characterized by a more-or-less simultaneous change in most, or all, of the cells, which is usually reversed when the environmental stress is removed, so it is not a permanently inherited trait. Some examples of this adaptive, phenotypic resistance are quite specific to particular organisms and antibiotics, for instance *E. coli* exhibits greater resistance to aminoglycosides under anaerobic conditions. However, the more important examples are generic, so, for example, there is considerable evidence that bacteria in general tend to exhibit greater resistance to antibiotics and biocides when they are growing as biofilms than when growing as freely suspended (planktonic) cells. There are likely to be several factors contributing to this increased resistance, but the slower growth rate of biofilm cells compared to planktonic ones is generally accepted as crucial. A second major example of phenotypic resistance is the induced synthesis of β-lactamases in response to the presence of a β-lactam antibiotic. Bacteria responding in this way possess the necessary genes all the time, so it is not a genetic *change* that they are exhibiting; it is merely a protective mechanism that they can switch on when required and switch off to avoid wasting energy when the β-lactamase is no longer needed.

As Figure 13.3 illustrates, antibiotic resistance arising from a genotypic change in the organism may either result from mutations in the genes the organism already has, or by the organism receiving new, additional genetic information from other cells.

13.3.1 Resistance arising by mutation

The frequency with which mutants arise in nature is influenced by the size of the bacterial population and its reproduction rate. Even when growing in the body, bacteria can achieve concentrations in excess of 10^9 colony-forming units per mL of body fluid and this, combined with doubling times of less than one hour and the fact that bacteria are usually haploid (so any mutation that occurs will be expressed rather than masked by a dominant homologous chromosome), means that mutation is a realistic and clinically important source of antibiotic resistance. One of the earliest examples was streptomycin resistance in *Mycobacterium tuberculosis*, which was reported shortly after the introduction of streptomycin in the late 1940s. Tuberculosis is a relatively difficult infection to eradicate so antibiotic treatment is required for several months, and it is this long period of exposure that gives the organism a protracted opportunity to mutate. The increased risk of resistance arising is similarly seen in other infections with a long period of antibiotic therapy, for example chronic *Pseudomonas aeruginosa* lung infections in cystic fibrosis patients. Resistance by enterococci to both vancomycin and aminoglycosides are more recent consequences of mutation.

Bacterial species differ in their potential to become resistant by mutation and some antibiotics are more vulnerable to resistance development by this means than others – fusidic acid and rifampicin are two which seem more susceptible than most. A mutation rate is defined as the probability of a mutant arising during a cell division cycle, so, for example, if there were 2×10^7 bacteria at the infection site and just one mutant arose after they had all divided, the mutation rate would be $1 \div 2 \times 10^7$, which equals 5×10^{-8}. Such values are sometimes quoted as if they were an intrinsic property of a drug/organism combination, whereas, in fact, they are influenced both by the concentration of antibiotic to which the bacteria are exposed and by the environmental conditions. Nevertheless, it is generally accepted that mutation rates to antibiotic resistance are typically 10^{-7} to 10^{-9}, but values as high as 10^{-5} and as low as 10^{-10} may occur. With bacterial concentrations at the infection site as high as 10^8 or 10^9 CFU/mL it is clear that tens or hundreds of mutants could arise per cell division, so resistance is a real threat. However, if two different antibiotics are used simultaneously, the chance of a single cell becoming resistant to both at the same time is the product of their mutation rates, not the sum. In other words, if the mutation rate for each antibiotic was 10^{-7}, the chance of a double mutant arising is 10^{-14}; this is the logic behind the use of multiple antibiotics being used to treat infections requiring a long period of therapy, such as tuberculosis, leprosy and HIV/AIDS (Chapter 9).

13.3.2 Resistance arising by receipt of new genetic information

Essentially, there are three ways in which new genes may be acquired by a bacterial cell.

13.3.2.1 Transformation

This describes the situation where a cell dies, disintegrates and releases its genetic information (DNA), which is picked up by another healthy cell that then transports it through its cell membrane and incorporates it into its own chromosome; it is often described as the transfer of 'naked' DNA. The process occurs naturally in only a few bacterial genera, for example *Streptococcus*, *Neisseria*, *Helicobacter* and *Acinetobacter*, and even in these it operates only under specific environmental or nutritional conditions. Organisms that do it naturally are described as competent, and for these species it is a realistic mechanism of acquiring resistance. It is less common and, consequently, less important than other mechanisms, but is very effective in some bacteria-antibiotic combinations such as benzylpenicillin or amoxicillin (and ampicillin) resistance in *Streptococcus pneumoniae* and gonococci.

13.3.2.2 Transduction

Here, a bacterial cell is infected by a bacteriophage (phage) which does not immediately kill it but becomes dormant inside the bacterial host (referred to as lysogeny) and its nucleic acid becomes incorporated into that of the host. The lysogenic state does not last indefinitely and may cease when the cell is exposed to an appropriate stimulus (for example chemicals or sublethal ultraviolet radiation); this causes the phage to be replicated, leading to death and lysis of the bacterial cell. The new phage particles are released and are free to infect new hosts, but they may carry with them small sections of bacterial DNA including antibiotic resistance genes from the original cell, and these may ultimately become incorporated into the chromosome of the new host. The bacteriophage acts, therefore, as a vector for the transmission of genetic information. Transduction is species-specific so it is not a mechanism whereby resistance can be transmitted between dissimilar bacteria, and probably the most

important example is that of β-lactamase genes transferred between strains of *Staphylococcus aureus*. It is worth noting here that transduction is important in other respects too: it is, for example, the means by which virulence factors, such as genes for toxin production, may be transferred from one cell to another.

13.3.2.3 Conjugation

Conjugation is the most important mechanism of horizontal gene transfer. Plasmids can be regarded as mini-chromosomes (typically about 1% of the size of the main bacterial chromosome) which contain genes that are not essential for normal growth and replication of the cell but which may confer an advantage to it under certain environmental conditions. The ability to produce toxins, fimbriae (to facilitate attachment of the cell to surfaces), and antibiotic resistance are just three of many characteristics determined by plasmid-encoded genes. Plasmids are replicated independently of the main chromosome and so a cell may contain many copies. Plasmids can be passed from one cell to another by a process of conjugation in which the donor and recipient cells are temporarily attached to each other by means of a sex (conjugation or F – for fertility) pilus (Figure 13.4). The sex pilus possessed by the donor cell acts like a lasso and locates the recipient then retracts to draw the two cells sufficiently close to each other to permit membrane fusion. The plasmid(s) inside the donor cell are replicated and a copy is passed to the recipient cell. In addition to the genes conferring antibiotic resistance, plasmids usually possess the genetic information to enable the cell

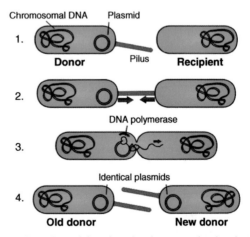

Figure 13.4 Bacterial conjugation (not to scale: the plasmid is, in reality, much smaller than the chromosome). *Source:* Adapted from http://commons.wikimedia.org/wiki/File:Conjugation.svg.

containing them to make the sex pilus itself, in which case they are termed self-transmissible plasmids. Non-self-transmissible plasmids have to rely on the formation of a sex pilus by other, self-transmissible, plasmids present in the same bacterial cell.

Two important features of horizontal gene transfer by conjugation that distinguish it from transformation and transduction are that the donor cell does not die in the process and there is a much lower degree of species specificity. This means that conjugation can occur not merely between the same strains of a species or between different species of a genus, but between genera, so antibiotic resistance genes can be transferred between, say, the various gut bacteria like *E. coli* and species of *Salmonella, Shigella, Proteus* and *Klebsiella* with relative ease. A consequence of this is that even bacteria that are regarded as nonpathogenic may represent a problem if they are harbouring antibiotic resistance plasmids. This is because resistance can be transmitted to pathogens following mating between the bacteria in the body (particularly in the colon where bacterial concentrations in excess of 10^{10} CFU/mL can occur, so cells are close together and collisions are relatively common). A further characteristic of plasmid-transmissible resistance is that the plasmids often carry genes conferring resistance to two or more dissimilar antibiotics at the same time. Such a plasmid is only of benefit to the bacterial cell when it is being exposed to one of the antibiotics: in an antibiotic-free environment the plasmid may even be a disadvantage to the cell because it has to expend energy making something that is of no value. Consequently it is often found that the plasmids may be spontaneously lost if the relevant antibiotic is not used – which is the logic behind the antibiotic cycling strategies that are part of stewardship programmes (see Chapter 14). If genes for, say, three different antibiotics are contained on the plasmid, only one of the three needs to be in regular use in the hospital ward for the plasmid to be retained and the bacterium would still exhibit resistance to the other two.

Antibiotic resistance genes may also be transferred between plasmids and chromosomes within cells by means of mobile gene sequences called transposons. These are sometimes described as 'jumping genes' to explain their potential to move from one location within the cell's DNA to another. Like plasmids, they can also promote the transfer of genes *between* bacteria by conjugation, but they differ from plasmids in that the genes required to initiate the conjugation process are located on the chromosome of the cell.

13.4 Resistance mechanisms at the cellular level

The sections above describe how antibiotic resistance is acquired or transferred between bacteria but do not explain the changes occurring in the cell which actually permit it to withstand the antibiotic. There are four major mechanisms, two or more of which may operate simultaneously in a resistant cell (Figure 13.5):

- the production of enzymes that degrade the antibiotic;
- alterations to the permeability of the bacterial cell so that antibiotic cannot enter as easily;
- a modification to the target inside the cell so that the antibiotic does not bind to it;
- efflux pumps, which remove the antibiotic from the cell.

The production of β-lactamase is a commonly encountered mechanism and it is important because β-lactams are, by far, the most commonly used antibiotics; these enzymes are considered in more detail below. β-lactams, however, are not the only group of antibiotics vulnerable to inactivating enzymes: aminoglycosides may be phosphorylated, adenylated (addition of adenine) or acetylated; chloramphenicol and streptogramins (quinupristin/dalfopristin – Synercid™) are inactivated by acetyltransferases, and macrolides are esterified (although this is not a major resistance mechanism).

Reducing the rate at which antibiotic enters the cell is a common resistance strategy that is particularly effective in Gram-negative bacteria which possess porins in their outer membrane (Figure 13.6). Porins are water-filled protein molecules that span the membrane and act as a pore through which hydrophilic molecules can diffuse and thereby enter the cell more readily. Resistance to many β-lactams, fluoroquinolones and tetracyclines has been shown to arise by reducing the number or the pore size of these molecules. Permeability reduction is not exclusively related to porins however: changes in cell membranes in Gram-positive bacteria (most of which do not possess porins) and alterations in lipopolysaccharides in Gram-negative outer membranes have also been shown to promote resistance.

Many antibiotic targets are enzymes, but microscopically visible structures like ribosomes may also be the binding sites for antibiotics in the cell. Subtle alterations in the structure of the target can dramatically

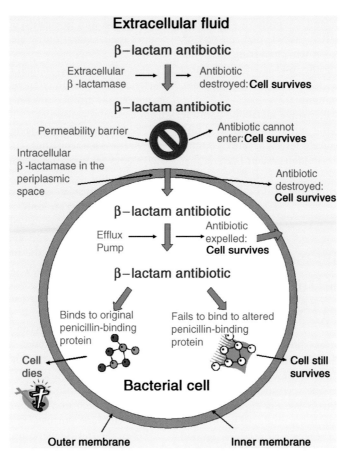

Figure 13.5 Diagrammatic representation of antibiotic resistance strategies and outcomes following exposure to a β-lactam antibiotic.

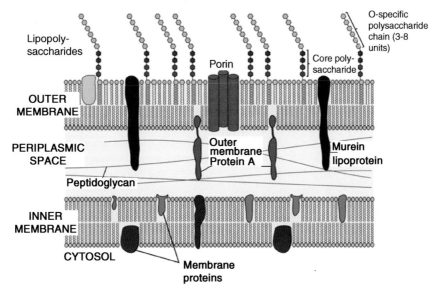

Figure 13.6 Structure of the Gram-negative bacterial envelope. *Source:* Adapted from http://commons.wikimedia.org/wiki/File: Gram-negative_Zellwand.svg.

reduce drug attachment and so promote resistance. This is the case with MRSA, which may possess an altered penicillin-binding protein conferring resistance to meticillin and flucloxacillin; and altered ribosomal RNA is a means of resistance to aminoglycosides, macrolides and streptogramins.

The cell adopting an alternative metabolic pathway that avoids the need for the antibiotic target is a fifth strategy, though it is relatively rare. The only example commonly cited is sulphonamide-resistant bacteria that do not need para-aminobenzoic acid to make folic acid; instead they utilize preformed folic acid in the same way as mammalian cells.

13.5 β-lactamases and extended-spectrum β-lactamases (ESBLs)

The first enzyme that was shown to hydrolyse the β-lactam ring of penicillin was called, not surprisingly, penicillinase, and that name continued to be used for many years when the enzyme was isolated from other bacteria, although it gradually became apparent that it was not just a single enzyme being produced by different species, but many different enzymes. Now it is recognized that there are literally hundreds of β-lactamases, which vary in terms of:

- Location in the cell: those from Gram-positive species are usually extracellular. The Gram negative enzymes are usually located within the cell in the periplasmic space, although extracellular β-lactamase activity may also occur.
- The location of the genes coding for the enzymes: either on plasmids or the chromosome.
- Their regulation: some β-lactamases are inducible (produced in response to the presence of the substrate) but many are constitutive (produced in lower concentrations all the time).
- Substrate specificity: some primarily hydrolyse penicillins, whereas other preferentially attack cephalosporins; some inactivate both, as well as other types of β-lactam antibiotics.
- Susceptibility to inhibitors: some β-lactamases are inhibited by, for example, clavulanic acid (Chapter 10) and tazobactam; others are not.
- Physicochemical properties, for example molecular weights vary from about 20 000D to more than twice

that value, and isoelectric points differ, which facilitates their separation by electrophoresis.

It is even possible, though not common, for a single bacterial species to simultaneously produce two different enzymes; one from genes on a plasmid and the other from chromosomal genes.

The location of the β-lactamase is one of the more significant characteristics. If it is an extracellular enzyme, as in *Staph aureus* for example, it is normally inducible and produced in large quantities; this is necessary because it is diluted by extracellular fluid. If, therefore, there are only a few cells present at the infection site, they may not produce sufficient enzyme to protect themselves and are therefore killed by the antibiotic. If, however, the infection is untreated and bacterial numbers increase, they may withstand the antibiotic because their enzyme-producing capacity is much greater; this is termed the 'inoculum effect'. Intracellular β-lactamase is normally located in the periplasmic space between the inner and outer Gram-negative membranes, and because it is not diluted by extracellular fluid it is present in much lower quantities, possibly just a few hundred molecules per cell, but that is sufficient to inactivate the antibiotic molecules, which may be entering slowly through the outer membrane porins (Figure 13.6).

When the third-generation cephalosporins containing an iminomethoxy group in their side chain were developed in the 1980s (for example, cefotaxime, ceftriaxone and ceftazidime) they were called extended-spectrum cephalosporins because they were active against a wider range of pathogens, particularly Gram-negative species, than the earlier ones. Initially they showed a good degree of resistance to β-lactamases and became widely used. Unfortunately, though, mutations in β-lactamase genes resulted in the emergence of new enzyme molecules with enhanced activity against the third-generation cephalosporins and these enzymes were termed extended-spectrum β-lactamases (ESBLs). They were first described in the mid-1980s – even as the third-generation cephalosporins were being introduced – and although their spread was slow to begin with, it gradually gathered pace so that now a significant proportion of strains of many Gram-negative pathogens including *E. coli*, *Klebsiella pneumoniae*, *Pseudomonas aeruginosa* and *Salmonella* species isolated in European and North American hospitals are found to be ESBL producers. This is a serious development because treatment options become very limited, the prospects of success much reduced, hospital stays extended and mortality is increased. Carbapenems are the drugs of choice for patients harbouring ESBL-producing pathogens, although even carbapenemases are becoming more frequent. Possibly the worst case scenario is an infection with a stably derepressed ESBL producer. The term 'stably derepressed' means that the repression-control mechanism that switches off production of an induced enzyme in the cell has ceased to work so that cell produces high levels of the β-lactamase all the time.

Acknowledgement

Chapter title image: http://commons.wikimedia.org/wiki/File:Antibiotic_disk_diffusion.jpg

Chapter 14
Antibiotic stewardship

UK national health service poster

KEY FACTS

- The development of new antibiotics has been declining steadily since 1980 because the pharmaceutical industry perceives anti-infectives to be less commercially viable than other drug categories.
- At the same time there has been a rise in resistance of several significant pathogens, with Gram-negative bacteria becoming particularly problematic in recent years.
- Antibiotic stewardship programmes have been developed with the aim of: restricting resistance development to preserve the antibiotics we currently have, reducing adverse effects, and improving both patient outcomes and the cost-effectiveness of antibiotic treatment.
- Hospital stewardship teams are normally multidisciplinary and they take responsibility for developing an antibiotics policy and a hospital formulary.
- One common approach for improving the prescribing and management of antibiotics is that of prospective audit with feedback; here prescribing decisions are reviewed and discussed in the light of information provided to prescribers.
- Alternatively, a more mandatory approach may be used whereby prescribers are restricted to a hospital formulary and may be required to gain authorization from senior colleagues to use antibiotics in restricted categories within it.
- Education of both patients and prescribers is a fundamental feature of most stewardship programmes.
- There has been a trend in the United Kingdom in recent years to make several (mostly topically applied) antimicrobial drugs available to the public without prescription.

Essential Microbiology for Pharmacy and Pharmaceutical Science, First Edition. Geoffrey Hanlon and Norman Hodges.
© 2013 John Wiley & Sons, Ltd. Published 2013 by John Wiley & Sons, Ltd.

14.1 Antibiotics: a resource to be protected

Alexander Fleming discovered the first antibiotic, penicillin, in 1928, but it was the need to treat infections of battlefield casualties in World War II that was the impetus for the drug's development and introduction into therapy in the early 1940s. Over the next 30 years many other antibiotics were discovered, and their use increased dramatically even though it was becoming apparent both that bacteria could become resistant to them and, significantly, the more the bacteria were exposed to an antibiotic, the greater the likelihood that resistance would arise. The pharmaceutical companies, however, continued to create new antibiotics to replace those whose value was eroded by resistance and, indeed, were so successful that in 1969 the US Surgeon General was quoted as saying that 'the time had come to close the book on infectious diseases'. Whilst there is some doubt now that he ever said it, the quote nevertheless epitomizes the complacency that prevailed at the time and the fact that it was taken for granted that a steady stream of new antibiotics would be produced. In the 1970s, however, the first outbreaks of MRSA occurred, and the prevalence of this and other so-called 'superbugs' rose dramatically over the next 20 years. By the turn of the century it was quite clear that there were several organisms for which there were few remaining effective antibiotics; one group in particular were named the 'ESKAPE' pathogens because of their ability to evade the commonly used drugs (*Enterococcus faecium*, *Staphylococcus aureus*, *Klebsiella* species, *Acinetobacter baumannii*, *Pseudomonas aeruginosa* and extended spectrum β-lactamase-producing strains of *E. coli* and *Enterobacter* species).

Unfortunately, this rise in resistance coincided with a change in attitude of the pharmaceutical industry towards antibiotics. Almost all of the chemical classes of antibiotics that we use today had been discovered by the 1980s, and the great majority of 'new' antibiotics introduced since then were not, in fact, entirely novel, but structural variations on existing drugs. Cross-resistance – the situation where bacteria exhibit resistance to a new drug because they have previously been exposed, and developed resistance, to similar drugs – is relatively common, so the best new antibiotics are those that have genuinely new chemical structures and mechanisms of action, but by 1980 these were becoming much harder to find. Consequently, several of the major companies abandoned research and development in antibiotics for one or more of the following reasons:

- The increasing difficulty and spiralling cost of developing new antibiotics.
- The fact that a course of treatment with a successful antibiotic is of such short duration – often just five days or less – that the financial return is far inferior to that on drugs to be taken for a lifetime, for example those to treat chronic conditions like hypertension, high cholesterol and diabetes.
- The recognition that it is the rule rather than the exception that bacteria will become progressively less sensitive to an antibiotic, so its therapeutic value (and sales) will diminish with time.
- The change in attitude from that where antibiotics were prescribed extensively and indiscriminately, to one where many clinicians and government expert committees advise that the few effective, new antibiotics that are produced – drugs like linezolid and daptomycin for instance – should be used sparingly to preserve their benefits; this further restricts the opportunity to recover development costs and make a profit.
- The difficulty of constructing FDA criteria-satisfying clinical trials for antibiotics to treat serious, potentially life-threatening, infections.

Figure 14.1 reveals the effect of this change in pharmaceutical industry direction on the number of new antibiotics licensed for use in the UK since 1979. It is this dramatic decline, together with the increase in antibiotic resistance and the rise of the 'superbugs' that has led to the realization both that some infections may literally soon become untreatable and that there is an urgent need to take steps to preserve the antibiotics we have now because replacements may not be forthcoming.

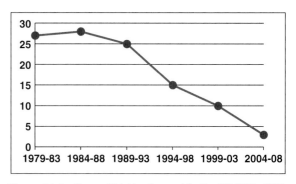

Figure 14.1   New antibiotics licensed in the UK 1979–2008. *Source:* MHRA.

It is the recognition that our existing antibiotics should not be squandered as a result of indiscriminate use but instead viewed as a precious resource to be passed on to future generations that has led to the concept of 'antibiotic stewardship'.

14.2 What is antibiotic stewardship?

The need to restrict antibiotic resistance was one of the major reasons for the introduction of so-called 'antibiotic policies' in many European and North American hospitals in the 1970s and 1980s. These policies focussed on the provision of advice and guidelines on appropriate antibiotic treatment for specific infections and the creation of hospital formularies that sought to control the use of antibiotics, particularly those considered to be of special value. The concept of antibiotic stewardship certainly incorporates these policies. Indeed, many people would regard them as the most fundamental, or even the only significant, component of a stewardship programme, but all of the following strategies could also be considered to be part of the stewardship concept because any means of reducing exposure of pathogens to antibiotics is likely to restrict resistance development, so they all have the potential to reduce consumption or improve the effectiveness of antibiotics:

- *Restricting the use of antibiotics in agriculture.* Most of the world's antibiotic production is used not to treat human infection but as a feed supplement for pigs, cattle and poultry. Adding antibiotics to animal feed has long been known to promote weight gain, so the practice was widespread in the twentieth century and by 2001 it was estimated that livestock accounted for as much as 70% of the total antibiotic consumption in the United States. Antibiotics that are used to treat human infection were banned as animal growth promoters in Europe in the 1990s, but they are still extensively used in this way in the United States and other countries.
- *Promoting more widespread use of vaccines.* Using vaccines more extensively is a means of reducing the number of infections and the need to treat them with antibiotics.
- *Improving infection control.* The term 'infection control' means the measures taken to prevent disease-causing organisms being passed from one person, or patient, to another, and it is widely acknowledged that improving the efficacy of these measures is the single most important step that can be taken to reduce

the incidence of nosocomial (hospital-acquired) infections and, consequently, the need for antibiotics to treat them.

- *Computer surveillance.* Monitoring the incidence of infections, antibiotic consumption and antibiotic resistance in both hospitals and the community provides the means to see how changes in antibiotic use influence resistance and the prevalence of selected pathogens. Heavy use of certain antibiotics can actually lead to increasing levels of some infections: for example, the chief risk factor for pseudomembranous colitis due to *Clostridium difficile* is previous exposure to third-generation cephalosporins or clindamycin. Heavy use of cephalosporins has been shown to increase the risk of vancomycin-resistant enterococci; and fluoroquinolone use has been associated with the prevalence of MRSA. Computer surveillance can help provide early warning of such problems and so pre-empt resistance development.
- *Better diagnosis by better laboratory testing.* It is desirable to treat an infection with an antibiotic that is specifically active against the particular pathogen responsible for the infection rather than a 'broad-spectrum' antibiotic that is effective against a wide range of organisms. This aim, however, can only be achieved when the pathogen is reliably identified and its antibiotic susceptibility profile determined in the laboratory, so any advance in the testing process that more reliably or more rapidly identifies the organism or its antibiotic susceptibility is likely to reduce the consumption of broad-spectrum antibiotics like amoxicillin or third-generation cephalosporins.
- *Early removal of catheters and cannulas from surgical patients.* Catheters and cannulas are flexible tubes that are inserted into the body to drain fluids from, or administer drugs to, a patient, and as a consequence they are an integral part of hospital treatment – particularly surgery. These tubes, however, necessarily breach the body's physical protection against infection by, for example, puncturing the skin and providing a route of entry for skin organisms. Catheters and cannulas are, therefore, vulnerable to colonization by bacteria which grow as a biofilm over their surfaces. These bacteria are particularly resistant to antibiotic treatment, partly because they are protected against the antibiotic by the polysaccharides and other polymers of which the biofilm is composed. Removal of catheters and cannulas at the earliest opportunity following surgery is another minor, but significant, practice that helps reduce the need for antibiotics and, as such, may be considered part of antibiotic stewardship.

14.3 The aims of stewardship programmes

Minimizing the development of antibiotic resistance, although of vital importance, is not the only aim of a stewardship programme, which, it is generally agreed, should:

- seek to improve patient outcomes;
- reduce the risk of adverse effects; and
- improve the cost-effectiveness of antibiotic treatment; as well as
- restrict antibiotic resistance development and help curtail the spread of antibiotic resistant organisms in the hospital and the wider community.

Improving patient outcomes means achieving the highest possible cure rates whilst minimizing the duration of treatment or stay in hospital. Clinical pathways (also called clinical guidelines, treatment guidelines, critical-care pathways or care maps) are now commonly used as part of a stewardship programme and they are simply a means of disseminating best practice by adopting standardized and sequenced procedures for the treatment of patients with specific infections such as tuberculosis or community-acquired pneumonia. In this way they seek to eliminate the variability in quality of care and patient outcomes that might arise when different medical teams each have their preferred antibiotic regimens. More sophisticated stewardship programmes might also incorporate computer surveillance of antimicrobial prescribing linked to individual patient records with the intentions of:

- ensuring that antibiotic dosing accounts for patient characteristics such as age, weight, renal and immune function; and
- of eliminating prescriptions for antibiotics to which patients have reported allergies or adverse events.

Whilst the majority of antibiotics are not particularly expensive drugs, they are widely used, especially in hospitals where, typically, a third of all patients receive them, and antibiotics may account for up to 30% of a hospital pharmacy budget. When these facts are considered together with the estimates that up to 50% of human antibiotic use could be eliminated without serious consequence it is clear that there is the potential to make substantial cost savings by adopting stewardship programmes that curtail antibiotic use. Whilst such programmes involve significant startup costs, these are likely to be recovered quickly since studies have shown decreases in antimicrobial use ranging between 22–36% and, in the United States, annual cost savings of $200 000–900 000 depending on hospital size. Not surprisingly, therefore, prescribing costs are one of the measures used to evaluate the benefits of a stewardship programme, but certainly not the only one. The levels of resistance to selected antibiotics, cases of hospital-acquired infection, and the length of time patients stay in hospital are all closely scrutinized to assess the success of stewardship strategies.

14.4 Hospital antibiotic stewardship programmes

There is no universally accepted definition of the term 'antibiotic stewardship' and no standardized structure for a hospital stewardship programme; they differ from one country to another because of the influence and advice of national committees and professional bodies, and their complexity is likely to be influenced by the size and resources of the hospital in which they are operated. Nevertheless, it is generally agreed that a stewardship team should be multidisciplinary and usually comprise:

- an infectious diseases physician;
- a clinical pharmacist with infectious diseases training;
- a medical microbiologist;
- an infection control professional;
- a hospital epidemiologist;
- an information technology specialist.

Their responsibilities would include deciding on the strategy of the stewardship programme and the preparation of both a written antibiotic policy and an antibiotic formulary. There is no single approach that is universally adopted for guiding and controlling antibiotic prescribing, but two core strategies that have been strongly advocated, particularly in the United States, are those of:

- prospective audit with intervention and feedback; and
- formulary restriction and preauthorization.

The word 'audit' in a clinical context means a review of procedures with the aim of identifying and adopting the best possible practices to improve patient care, so a prospective audit would be designed to review antimicrobial prescribing *as it happens* so that suggestions can

Figure 14.2 Educational posters used in the United States and United Kingdom to inform the public about the limitations of antibiotics and the threat of resistance. *Sources:* Get Smart poster, US Centers for Disease Control and Prevention; http://commons .wikimedia.org/wiki/File:CDC_Get_Smart_poster_healthy_adult.jpg (left); UK Department of Health poster. © Crown copyright 2008 (middle); UK Department of Health poster. © Crown copyright 2008 (right).

be made for improvement whilst treatment of the infection is still in progress. It has been shown that providing information to prescribers about the dosage, pharmacokinetics, resistance patterns, toxicity or other characteristics of the antibiotics they intend to use and, where necessary, the preferred alternative(s) can significantly improve the quality of treatment and reduce unnecessary prescriptions. Such a policy relies on education and persuasion of prescribers, but the alternative of formulary restriction and preauthorization is a more mandatory approach. Here, the recommended antibiotic treatment for specific well defined infections would be described in the hospital's clinical pathways and the drugs available would be listed in a hospital formulary.

Usually the stewardship team would evaluate the drugs to be routinely stocked on the basis of their efficacy, safety and cost. Antibiotics to be used with particular care, either because of their value as drugs of last resort or because of cost, may be placed in a restricted category in the formulary so that they may only be prescribed with the approval of senior clinicians. This approach of formulary restriction with preauthorization has been shown to produce immediate and significant reductions in antibiotic use and costs, but its benefit in restricting resistance development is less well defined. There are several common elements of stewardship strategies which are listed below; not all of them have been shown to be of unequivocal benefit however.

14.4.1 Education

Attempts to educate the public about antibiotics and their limitations are well established and are prominently supported by posters and information leaflets in surgeries and health centres (Figure 14.2). Educating prescribers about the relative merits of different antibiotics, however, is usually multifaceted and may employ inhouse teaching sessions, conferences and email alerts as well as updates to hospital antibiotic guidelines. Evidence has shown that each of these used individually has only a minimal or transient effect, and that they are far more beneficial when employed together, and yet more effective when used in combination with active strategies like audit and intervention.

14.4.2 Guidelines and clinical pathways

Clinical guidelines are being used with increasing frequency, particularly for well defined and easily diagnosed infections like pneumonia and cellulitis, and they have been shown to be effective in reducing antibiotic use and improving patient outcomes (as measured by shorter periods of intravenous antibiotic therapy and shorter hospital stays). Significantly, though, it has again been found that one strategy in isolation is not ideal, and that the adoption of clinical guidelines works best when implemented together with improvements both in diagnosis and in better planned nursing care and discharge management.

14.4.3 Antimicrobial order forms

These are particularly valuable where antibiotics are used as prophylactic cover to minimize the risk of infection during surgery. In the absence of order forms with an automatic stop date there is a tendency for therapy to be continued after the operation for an unnecessarily long period, in the latter part of which the antibiotic affords little or no benefit. The introduction of order forms with an automatic stop date (typically 1–2 days after surgery) beyond which further physician justification is required to continue treatment have been shown to reduce antibiotic consumption without adverse consequences for the patient.

14.4.4 De-escalation of therapy

Many infections are treated initially on an empirical basis. This means that the precise identity of the pathogen is unknown and broad-spectrum treatment is started using antibiotics that are normally effective against the organisms thought most likely to be responsible. Whilst this strategy maximizes the chance of successful early treatment, it has the disadvantage that excessive exposure to broad-spectrum antibiotics increases the chance of resistance development. It is usually the policy therefore, to switch to a more specific narrow-spectrum antibiotic once the causative agent is identified.

14.4.5 Parenteral to oral conversion

Antibiotic treatment for serious infections in hospital is usually initiated via the intravenous route in order rapidly to achieve high concentrations at the infection site. However, the use of injected antibiotics is largely confined to hospitals, so this limits their sales and leads to companies charging higher prices in order to recover development costs before patents expire. Consequently, there is a financial incentive to switch from injected to oral antibiotics, but, apart from that, intravenous lines are not popular with patients so substitution of oral therapy may facilitate earlier removal of a cannula which not only gains patient approval but reduces the risk of

antibiotic-resistant biofilm bacteria growing on the cannula itself.

14.4.6 Dose optimization

Whilst the above aspects of stewardship programmes focus primarily on the goals of reducing resistance and avoidance of unnecessary antibiotic use, dose optimization is very much concerned with improving patient outcomes. Here, the intention is to maximize both the safety of the antibiotic treatment and the likelihood of its success in eradicating the infection by tailoring the dosing according to the patient's individual characteristics including age, weight, immune status and renal function. The site of the infection and the ease with which effective antibiotic concentrations can be achieved may also modify the choice of drug and frequency of dosing. After administration, antibiotics usually do not achieve uniform concentrations in all body fluids: normally the concentration is significantly higher in the urine than in the blood, which, in turn, is often higher than that in, for example, the cerebrospinal fluid, prostatic fluid or bile. Consequently the site of the infection and the pharmacokinetics of the antibiotic may be important considerations when selecting the dose.

14.4.7 Antimicrobial cycling

There is abundant evidence that the more antibiotics are used, the greater the likelihood of resistance development, so it would be logical to expect that if a particular drug were taken out of use for a significant period of time this would result in a reduction, or, at worst, no further increase in the frequency of isolation of organisms resistant to it. On this basis 'cycling' of antibiotics has been introduced as part of antibiotic policies in some hospitals since the 1980s, but with variable results. There is certainly evidence that in intensive care units especially, the withdrawal of selected antibiotics, particularly those used to treat infections with Gram-negative bacteria, can have a beneficial effect, but data indicate that reintroduction of the drug in question may lead to a rapid return to the original resistance levels. The benefits of cycling of antibiotics are, in any case, often hard to quantify because it may be difficult to achieve total withdrawal of an antibiotic; experience has shown that 'rested' antibiotics are still prescribed because allergy or toxicity may preclude the use of the current 'in-use' alternative drug or, simply because of prescriber resistance to the policy.

14.4.8 Combination therapy

Using antibiotics in combination is another strategy that is of undeniable benefit in certain circumstances but there is insufficient evidence to support it as a routine means of limiting resistance development. The situation in which combination antibiotic therapy is most beneficial is when antibiotics need to be administered for a long period of time because this would provide the greatest opportunity for the infecting organism to mutate or acquire the genetic information conferring resistance. Thus, emergence of resistance can be delayed or eliminated by using two or more antibiotics together in, for example, tuberculosis which typically requires six months treatment, chronic lung infections of cystic fibrosis sufferers (treated intermittently for months or years) and in HIV/AIDS patients where the treatment lasts for a lifetime. Combination therapy may also be appropriate during early empirical treatment of an infection before the causative organism is identified, but with these exceptions it is now rarely encountered as a routine element of a stewardship programme.

14.5 Availability of antibiotics to the public

The ease with which members of the public can obtain antibiotics varies greatly throughout the world. In many countries antibiotics can simply be bought over the counter at a pharmacy, but in other countries, particularly those of the European Union, antibiotic availability is more rigidly controlled and they should, in theory, only be supplied on prescription. However, the philosophy of restricting antibiotic availability in order to restrict the emergence of resistance is under threat from several quarters. Buying antibiotics via the Internet is an obvious problem, as is the practice recently cited by the Infectious Diseases Society of America in evidence to a US Congressional Committee where 'grocery stores and pharmacies give prescribed antibiotics away for free as a marketing ploy to lure customers into their stores'. In the United Kingdom there has been a trend since the early 1990s of deregulating selected antibiotics, particularly those used topically like clotrimazole, aciclovir and chloramphenicol, to make them available without prescription and, in some cases, without any medical or pharmaceutical advice, through supermarkets. Concern about the possible consequences in terms of resistance development has been fuelled more recently by the reclassification of azithromycin. In 2008 it became available for chlamydia treatment without prescription from a medical practitioner and, as such, it was the first oral product to be deregulated, a move which was followed by applications for controls on trimethoprim and nitrofurantoin to be similarly relaxed to permit self-medication of urinary-tract infections. These events have attracted adverse reaction from several sections of the medical community; both the UK Chief Medical Officer and the Society for Antimicrobial Chemotherapy have cautioned against any further relaxation of antibiotic controls that would facilitate their unsupervised or unnecessary use.

Acknowledgement

Chapter title image: UK Department of Health poster. © Crown copyright

Part III

Microorganisms and the manufacture of medicines

Chapter 15

Bioburdens: counting, detecting and identifying microorganisms

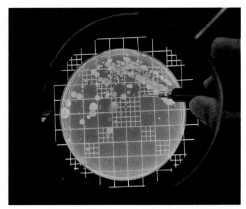
Counting bacterial colonies

KEY FACTS

- The term *bioburden* means the number and type of microorganisms present in or on a pharmaceutical raw material, medicine or medical device.
- The pharmacopoeias describe procedures both for counting microorganisms and for confirming the absence of named 'objectionable organisms'.
- These procedures may be based upon traditional cultivation methods or may use automated instruments that rely on other technologies like impedance, microcalorimetry or fluorescence; traditional methods are still, by far, the most common.
- For different categories of nonsterile medicine there are specifications in the pharmacopoeias for the maximum permitted levels of bacteria and fungi, together with requirements for the absence of one or more of the major objectionable organisms: *E. coli*, *Salmonella* species, *Staphylococcus aureus*, *Pseudomonas aeruginosa* and *Candida albicans*.
- Commercially available test kits and automated instruments are available for identifying different categories of bacteria and yeasts.
- The identification of moulds is based largely on microscopy and is usually undertaken in specialist centres.

Essential Microbiology for Pharmacy and Pharmaceutical Science, First Edition. Geoffrey Hanlon and Norman Hodges.
© 2013 John Wiley & Sons, Ltd. Published 2013 by John Wiley & Sons, Ltd.

15.1 What does 'bioburden' mean?

The word 'bioburden' means different things to different people. It is normally used to mean the number, or concentration, of microorganisms present in, or on, a product. Thus, the number of organisms in, say, a tablet, or on the surface of a medical device (for example, a catheter, cannula, or surgical implant) would be a bioburden, as would their concentration in a raw material like water or a medicine like a cream or ointment. More correctly though, the United States Food and Drugs Administration (FDA) and the European Union Guidance on Good Manufacturing Practice define a bioburden as 'the level *and type* of microorganisms that can be present . . . ', which implies the need to identify the organisms or, at the very least, confirm that particular objectionable organisms are absent.

This prompts the question of what is meant by 'objectionable'. An organism may be described in this way for one or both of the following reasons:

- because it is a pathogen so its presence in a medicine may cause an infection; or
- its presence may be indicative of poor-quality raw materials or poor manufacturing procedures.

Both *E. coli* and *Staphylococcus aureus*, for example, are pathogens, so they should not be found in medicines because they are a health hazard anyway. But apart from that, *E. coli* is an organism normally found in the colon of mammals so its presence is suggestive of faecal contamination, and *Staph aureus* commonly arises on the skin, so its presence in a medicine may implicate the manufacturing personnel as a source of microbial contamination. The pharmacopoeias therefore describe tests to detect these organisms, although the tests are not applied indiscriminately to all products but specifically to those where the presence of the organism is a realistic possibility or where it may represent a particular risk of infection because the product is applied to a vulnerable site in the body. Gelatin, for example, is made from animal skin and bones so it has its origins in the slaughterhouse where contamination from animal faeces may arise; it is therefore subject to tests for the absence of both *E. coli* and *Salmonella*. The pharmacopoeias also describe tests for the absence of *Pseudomonas aeruginosa* and *Candida albicans*, but the term 'objectionable organism' should not be thought of as applying exclusively to the five organisms named here; other organisms could be considered objectionable in particular circumstances or products.

A quantitative bioburden determination simply means a count of the organisms present, and the term often used is a total viable count (TVC). There is scope for confusion here though, because a distinction is sometimes made between a viable count, meaning just living organisms, and a total count, meaning both living and dead. The situation is further complicated by the fact that there are different categories of TVC: the European Pharmacopoeia makes a distinction between total aerobic microbial count (because anaerobes are usually disregarded) and a total yeast and mould count (which would be performed using different culture media); it also describes a counting method for enterobacteria (*E. coli* and related organisms).

Apart from environmental monitoring (see Chapter 17), bioburden determinations are the most common microbiological test undertaken in the pharmaceutical industry. They are performed not only on raw materials and finished manufactured products, but possibly at various stages during the manufacturing process too – for example, immediately prior to heat sterilization in order to confirm that the microbial load to be killed is within the capacity of the sterilization process in question. Even in sterile medicines where all the microorganisms are ultimately killed, it is important to keep a check on the levels of contaminants entering the product as it is made; this is because bacteria remain pyrogenic (cause fever when injected) when they die, so a sterile product can still fail a pharmacopoeial pyrogen or endotoxin test. The requirement for bioburden monitoring is not restricted to large companies with extensive manufacturing operations; a small company that simply packs tablets that have been made elsewhere would be expected to have records of the bioburden in water used for cleaning the packing equipment. Both for this reason and because it is the most widely used ingredient of manufactured medicines, water is subjected to bioburden testing more commonly than any other material.

15.2 Traditional counting methods

There are several methods available for measuring the TVC using both traditional cultivation procedures and modern automated instrumentation. The traditional approach involves placing a sample of the material to be tested onto, or into, gelled culture medium in a Petri dish (plate) and counting the visible colonies that arise after incubation. A single colony may develop from an individual cell or from a group of cells attached or

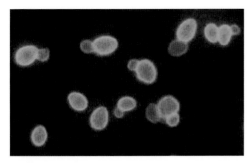

Figure 15.1 Stained yeast cells. *Source:* PHIL ID #291; Photo Credit: Maxine Jalbert and Dr. Leo Kaufman, Centers for Disease Control and Prevention.

'clumped' together, so the count is not expressed as *cells* per ml or gram but as colony forming units (CFU or cfu – there is no convention for use of upper or lower case letters). Figure 15.1 for example shows a total of 19 yeast cells, but these would give only 10 colonies because those in pairs or trios would, in each case, produce just one colony.

The United States Pharmacopoeia states that Petri dishes containing between 25–250 colonies produce the most reliable bioburden results. Mean colony counts lower than 25 have unacceptably poor precision, whereas counts higher than 250 might produce an artificially low result either because of overcrowding where two colonies very close together merge to be counted as one, or because there could be competition for nutrients so that some colonies may not grow large enough to be seen. In reality, though, colony counts of hundreds rarely arise; indeed, even colony counts in the tens are uncommon because most raw materials are now made using chemical processes involving high temperatures, extremes of pH and organic solvents, all of which kill microorganisms. It is usually only raw materials of 'natural' origin that have high bioburdens: mined minerals like talc, kaolin and bentonite have contaminants from the soil; so, too, have materials of vegetable origin such as starches, gums and thickening agents. Apart from gelatin, raw materials of animal origin are not common.

If the material to be tested was expected to have a high bioburden it would be necessary to dilute it in order to achieve a 'countable' number of colonies on the plate, but water, and raw materials that were expected to have a low count as a result of their manufacturing process, would require no dilution. If there were no historical data to indicate the degree of dilution required it would be necessary to perform a series of dilutions (typically tenfold increments) and place a known volume of each dilution into separate Petri dishes. Only those plates

having counts in the 'reliable' range would be used; the others containing more than 250 or fewer than 25 colonies would be discarded without counting. If, however, the only result available was that from a plate containing fewer than 25 colonies, that value would be recorded. It is better to have a result with a relatively high percentage error than no result at all!

15.2.1 Pour plates

The most common method for determining the TVC is the so-called pour-plate method (Figure 15.2). If the material to be tested is an aqueous liquid, a known volume is simply placed in the base of the Petri dish and 10–25 ml of molten culture medium (typically tryptone soya agar or plate count agar at 45°) is poured onto it and quickly mixed by gentle swirling, then the plate is placed on one side for the agar to set.

If the sample is a solid that is soluble in water that solid would normally be dissolved, but if it were insoluble then a suspension would be used. The problem in the last case would be to ensure that the suspension remained uniformly dispersed during pipetting and any dilution steps. Despite the fact that bacteria possess a cell wall that should provide osmotic stability, some species may suffer a degree of viability loss if diluted quickly in cold water; consequently, isotonic buffer (phosphate buffered saline for example) or peptone water are used as solvents or diluents. In the case of peptone water, however, there is a risk of the contaminating bacteria growing to give an artificially high count if a long time is required to dissolve the sample.

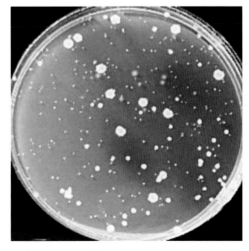

Figure 15.2 A pour plate.

Because the sample is dispersed through the medium before the gel sets, the colonies that grow are similarly dispersed throughout the agar. Consequently the colonies are usually of different sizes (Table 15.2) because there is less oxygen available within the gel than at its surface and most organisms are either strict aerobes or facultative anaerobes (see Chapter 2) which will grow best where the oxygen concentration is highest. *Bacillus* species, for example, are relatively common contaminants in bioburden tests and most are strict aerobes, so the possibility exists that colonies at the bottom of the agar might be too small to be counted.

15.2.2 Spread plates

The problem of small colonies due to lack of air may be overcome using the alternative counting method which is the surface-spread method (Figure 15.3) in which the sample is literally spread all over the agar surface with a sterile glass or plastic 'hockey-stick' spreader then allowed to soak into the gel. A variation on this procedure is the Miles–Misra surface drop method (Figure 15.4) where discrete drops of suspension are placed on the surface to give clusters of colonies in the locations where the drops soaked into the gel.

15.2.3 Membrane filter method

Both pour plate and spread plate methods are really only suitable for materials with relatively high bioburdens. The pour plate method typically uses a 1.0 ml sample, but the surface spread method normally uses 0.1–0.5 ml because larger volumes will not soak into the gel (and if they don't

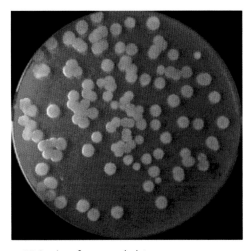

Figure 15.3   A surface-spread plate.

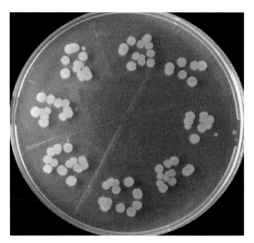

Figure 15.4   Miles–Misra surface-drop method.

soak in, the bacteria, rather than growing as separate colonies, grow and swim in a 'lake' of liquid on the surface and completely cover it in one large colony). In order to have, say, 30 colonies on a surface spread plate therefore, there must be a minimum concentration of 60 CFU/ml in the sample (30 CFU in 0.5 ml). However, many pharmaceutical materials normally have lower levels than this and Purified Water EP is *required* to have no more than 100 CFU/ml simply to meet the pharmacopoeial standard. In this situation, therefore, the last commonly used bioburden method of membrane filtration is normally employed. Here, the sample (which, in the case of water, may be a very large volume of 100 ml or more) is passed under vacuum through a sterile filter membrane having a pore size sufficiently small to retain all contaminating organisms on its surface (usually 0.45 μm). The membrane is then placed (without inversion) onto the agar and, assuming no air bubbles have been trapped underneath, the nutrients from the gel diffuse up and allow the bacteria to grow on the surface of the membrane just as if they were growing on the gel itself (Figure 15.5).

This method affords two other advantages:

- It can be used to determine the concentration of organisms in oils, or even in ointments which can be dispersed in oil; it is necessary, though, to use membranes that are oil-resistant – the cellulose acetate or cellulose nitrate membranes normally used may not be satisfactory – and to ensure an adequate flow rate by using a low viscosity oil or gentle heat – typically 40 °C.
- Filtration is unquestionably the best method if the sample is likely to contain antimicrobial chemicals. Antibiotic products or those containing preservatives

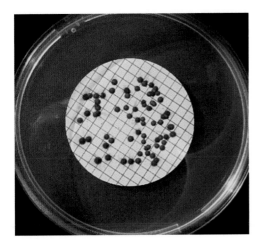

Figure 15.5   Membrane filter method.

(multidose eye drops and injections for example) are not self-sterilizing, so microorganisms may survive in them without necessarily growing. If, however, the sample was diluted and counted by a pour-plate or spread-plate method, the carry-over of preservative onto the plate may still be sufficient to inhibit the development of colonies during the standard incubation period, and this may result in an artificially low count. When membrane filter methods are used for samples such as these, it is normal for the membrane to be rinsed with peptone water (which may even contain a specific inactivator for the chemical in question – see Table 15.1) and so any contaminating organisms have residual preservative washed away and are thus free to grow normally.

The pharmacopoeias describe validation procedures to confirm that the counting methods are sufficiently sensitive to detect low levels of contamination and that any preservatives are effectively neutralized.

15.2.4 Most probable number method

There is one other counting procedure available, the most probable number method (MPN), which is only used when the methods described above are unsuitable – typically when the sample contains insoluble material that would interfere with plate counts, as it would in many herbal products for example. An MPN count involves adding different known weights or volumes of sample to tubes of liquid media, which are incubated, then examined for the presence or absence of growth. If the contamination level is sufficiently low, some of the tubes will receive a sample that does not contain any contaminants at all and will therefore show no signs of microbial growth after incubation. Other tubes, which receive a sample that does contain contaminants, will become turbid (cloudy) after incubation. There are different experimental designs available, but typically three tubes each receive 0.1 ml or 0.1 g sample, whilst three more receive 0.01 ml or g, and the last three 0.001 ml or g. The numbers of tubes in each set of three that show growth are recorded and the most probable number of contaminants read from a table of possible results which is constructed on the basis of statistics. The method is best suited to samples containing 10–1000 CFU/ml or g; higher or lower levels of contamination are outside the range of the method.

Table 15.2 identifies the important advantages and disadvantages of the counting procedures mentioned.

15.2.5 Calculation of concentration of microorganisms in a sample

At the end of a viable counting procedure we should have colonies growing on agar plates at a variety of different dilutions. Some of these will be overcrowded as the dilution was not sufficient, while others will contain too few

Table 15.1   Inactivators (neutralizers) for common antimicrobial substances.

Antimicrobial agent[a]	Inactivator
Quaternary ammonium compounds, parabens and chlorhexidine	Lecithin with or without Polysorbate (Tween) 80
Thiomersal and other mercurials	Sodium thioglycollate
Halogens	Sodium thiosulphate
Aldehydes	Glycine or sodium bisulphite
Beta-lactam antibiotics	Beta-lactamase

[a]Phenols, alcohols and weak acid preservatives may be inactivated by dilution alone.

Table 15.2   Advantages and disadvantages of traditional bioburden determination methods.

Counting method	Advantages	Disadvantages
Pour-plate	• Easy to undertake • Will detect lower concentrations than surface spread method because of the larger sample volume	• The use of relatively hot agar carries the risk of killing some sensitive contaminants, so giving a low result • Small colonies may be overlooked
Surface spread plate (and Miles–Misra method)	• Produces large colonies that retain their characteristic morphology, so aiding identification in some cases	• Least sensitive (i.e. least suitable for low bioburden samples) • If the liquid does not soak into the gel no discrete colonies are formed, so no results available, therefore plates require predrying (also called 'overdrying')
Membrane filtration	• Suitable for all bioburden levels • May be adapted to measure bioburdens in oily materials using nonaqueous solvents • Antimicrobial substances that may be in the sample are physically separated from surviving organisms and do not slow their growth into colonies	• Relatively more time-consuming and expensive of materials • Viscous liquids will not easily pass through the membrane and it may become blocked with suspended particles
Most probable number method	• May be suitable for samples that cannot be counted by other methods – particularly water-insoluble materials and herbal medicines	• Laborious and expensive in terms of materials, glassware and incubator space • Relatively large margin of error

colonies as the dilution was too great. Hopefully, there will be some plates which contain an acceptable number of colonies to give a valid result (25 to 250 colonies). In order to give statistically valid results it would normally be necessary to plate out each dilution in triplicate.

Calculation of the contamination level in the original sample requires knowledge of three things:

• the mean colony count;
• the volume of dilution placed on or in the agar;
• the extent to which the sample was diluted before plating.

As an example, let us say we examined an aqueous sample by performing six tenfold serial dilutions and then carried out the spread plate technique spreading 200 μl (0.2 ml) on each of three plates per dilution. The results obtained might look something like those presented in Table 15.3.

From this table it is evident that dilutions A to C are of no value because the colony counts are far too high. Similarly, dilutions E and F give counts which are much too low. The only dilution which gives reliable results is dilution D and so we can ignore all the others and just use that one.

Table 15.3   Specimen results from a viable count.

Dilution	Dilution factor	Colony count 1	Colony count 2	Colony count 3	Mean colony count
A	10^1	TNTC[a]	TNTC	TNTC	TNTC
B	10^2	TNTC	TNTC	TNTC	TNTC
C	10^3	453	521	419	464
D	10^4	85	79	81	82
E	10^5	7	6	8	7
F	10^6	0	1	0	<1

[a]TNTC = too numerous to count

This result tells us that the mean colony count obtained from spreading 0.2 ml onto the surface of an agar plate was 82. Since we need the result to be in colony forming units per *millilitre* we must multiple this value by 5. If we had spread 0.1 ml on the plates then we would have to multiply by 10. For our results this gives us a value of 410 – in other words dilution D contained 410 colony forming units per ml. However, this is still not the final answer because we want to know the concentration of cells in the original sample. In order to get dilution D the original sample was diluted by four tenfold serial dilutions – that is, by a factor of 10^4 or 10 000. To get the final result we therefore have to multiply our value of 410 by 10^4, which gives us a concentration in the original sample of 4.1×10^6 CFU/ml.

A general formula which is applicable to all methods of viable counting is as follows:

$$\text{Viable count of original sample} = \frac{\text{Mean colony count}}{\text{Volume of dilution used}} \times \text{dilution factor}$$

15.3 Detection of objectionable organisms

Injections, eye drops and certain other types of medicines are sterile, but the great majority of dosage forms are nonsterile and therefore, by definition, may contain some microorganisms. The quality of nonsterile products is controlled by the pharmacopoeias in two ways:

- there are limits on the total number, or concentration, of organisms that may be present;
- particular objectionable organisms must be absent in a specified weight of material; so, for example, in EP-quality gelatin *Salmonella* should be absent in a 10 g sample and *E. coli* absent in 1 g sample.

The word 'absent' here should be viewed with the same caution as the word 'sterile' used in the context of a sterility test. The fact that a product passes a pharmacopoeial sterility test does not *guarantee* it to be sterile, nor is a product that passes an EP detection test for *E. coli* guaranteed to be free of this organism. In each case there is a possibility that testing a larger sample or the use of different methods would reveal organisms that are not detected by the EP test. Nevertheless, the EP describes detection tests for both *E. coli* and *Salmonella* and for *Pseudomonas aeruginosa*, *Staphylococcus aureus*, and *Clostridium perfringens*. In EP 6.0 (2008), there are two

sets of tests described: those which are currently applicable in the EP, and the newer harmonized versions (which are identical to the United States and Japanese Pharmacopoeias) to be implemented when the individual product monographs are updated. It should be emphasized that these are not identification methods; they provide no information about the identities of other organisms that may be present (which may even include other pathogens) but merely seek to confirm that particular named organisms are absent.

The details of the procedures are too complex and lengthy to describe here, but for each organism the same steps apply:

1. Dissolution or dispersal of the sample in a suitable liquid culture medium and, where necessary, inactivation of any substances that might inhibit the growth of the organism under test.
2. Enrichment: this means increasing the relative concentration of the test organism by growing in a liquid medium that inhibits other contaminants but allows free multiplication of the organism of interest.
3. Streaking liquid cultures (see Chapter 2) from step 2 onto selective agar media that usually permit easy recognition of any colonies of the test organism that might arise.
4. The use of specific biochemical or immunological confirmatory tests, often using miniaturized test kits.

The selective media currently recommended in the pharmacopoeias for the four principal objectionable bacteria are:

- MacConkey's agar for *E. coli*.
- XLD agar for *Salmonella* species (previously brilliant green agar and deoxycholate citrate agar were also used).
- Mannitol salt agar for *Staphylococcus aureus* (previously Baird–Parker agar was also used).
- Cetrimide agar for *Pseudomonas aeruginosa*.

Again, there are validation procedures that seek to confirm that the detection tests will actually reveal the presence of low concentrations of the specified organisms when they are present in a sample together with higher concentrations of other species that might mask their presence. In reality the detection of objectionable organisms is sometimes less straightforward than anticipated for several reasons:

- In most species of bacteria, atypical strains arise that do not conform to the standard textbook descriptions; for

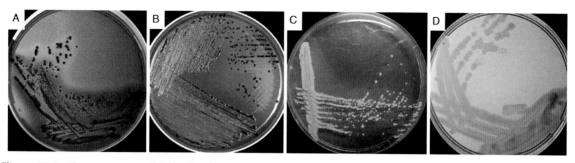

Figure 15.6 The appearance of 'objectionable organisms' on selective media. A. *E. coli* on MacConkey's agar, showing characteristic red colonies and precipitation of bile as a result of acid production; B. *Salmonella* species on XLD agar, showing alkali and hydrogen sulphide production indicated by pink agar and black precipitate in the colonies respectively; C. *Staphylococcus aureus* on mannitol salt agar showing yellow colonies that have produced acid as a result of mannitol fermentation; D. *Pseudomonas aeruginosa* on cetrimide agar, showing the characteristic green pigmentation.

instance, most strains of *Ps. aeruginosa* produce blue or green pigments, but a few strains are orange/brown and some are colourless.

- The selective agar media are not absolutely specific for the organism of interest, so other organisms having a similar appearance will grow and cause confusion. For example, some *Proteus* species will grow on both Baird–Parker medium and mannitol salt agar which are commonly used for the detection of *Staph. aureus*.
- The pharmacopoeial descriptions of the 'target' organisms on the recommended culture media are sometimes brief and uninformative.
- The recommended incubation conditions are imprecise with respect both to time and temperature and the appearance of the colonies may vary depending on the precise conditions used.

Despite these shortcomings, the objectionable organisms can usually be identified as characteristic colonies on selective media as shown in Figure 15.6.

15.4 Automated bioburden determinations

The traditional methods of bioburden testing have been used for many years and the basic principles upon which they operate have changed little since the early days of microbiology two or more centuries ago. The methods work well and they are still very widely used, but their main drawbacks are that they are labour-intensive and time-consuming. The incubation period for a total yeast and mould count can be five days or more, which means

that a raw material cannot be used, or a manufactured batch of medicine cannot be released for sale, until that incubation period is complete and the results recorded. There is a problem, too, in that a degree of experience is required to recognize colonies of particular organisms and interpret the results of detection tests. The need for more rapid results has been the primary driver for the development of a variety of automated and instrumented methods for the detection and enumeration of microorganisms, and although these methods are often more suited to the food and dairy industries, some of them have been applied to pharmaceuticals where they are slowly gaining regulatory acceptance.

The European Pharmacopoeia general text section 5.1.6 describes the operating principles and applications of 14 such methods together with information on the critical aspects of each method that might influence the reliability of the results. The bases of the methods are quite varied; they include, for example:

- Measurement of changes in electrical conductance and heat generation in a culture medium as organisms grow (impedance methods and microcalorimetry respectively).
- Methods that measure ATP concentrations (which are directly proportional to cell numbers).
- Procedures using fluorescent dyes that only stain living cells.

It is not the intention to provide an account of all the methods here, particularly because several of them have limited potential applications – only detecting the presence or absence of living organisms in general rather than the numbers or the species, for example – and very few have actually been used yet in the pharmaceutical industry, but flow cytometry and nucleic acid amplification techniques (the polymerase chain reaction – PCR) warrant further explanation.

Flow cytometry can be used to count the number of living cells in a liquid sample, but it does not identify them in any way. The sample is treated with a fluorophore – a chemical which, initially, is nonfluorescent but which is converted by the enzymes inside living cells to a fluorescent derivative – which enables the cells containing it to be detected as the sample is passed via a fluorescence microscope linked to a computer. This method is particularly suitable for liquids like pharmaceutical grades of water which have both low viscosity and low levels of suspended solids which might naturally fluoresce and be incorrectly recorded as bacteria.

The polymerase chain reaction (PCR), which is extensively used in forensic science for 'genetic fingerprinting' and the identification of criminal suspects, has the potential to be adapted for the rapid detection of objectionable organisms because it can be so specific and sensitive. On the other hand, though, PCR methods are susceptible to errors due to cross-contamination from previously analysed samples and the effects of substances in the sample which might inhibit the activity of the enzymes involved.

When new methods are being considered for adoption as replacements for traditional ones the regulatory agencies (the Food and Drugs Administration (FDA) in the United States and the Medicines and Healthcare products Regulatory Agency (MHRA) in the United Kingdom, for example) always require the new method to be fully characterized and shown to be at least as reliable as the one being replaced before approval is given. In addition to high capital equipment and reagent costs, this validation process can be a significant disincentive for the development and adoption of new methods in the pharmaceutical industry. It is necessary to provide data on:

- the accuracy and precision of the new method (which do not mean the same – accuracy is how close the result is to the true value whereas precision measures how much results vary when the same sample is assayed repeatedly);
- the method's detection limit and operating range,
- specificity (whether it will incorrectly record other objects or species as the organism of interest);
- linearity (whether the instrument reading is directly proportional to the concentration of bacteria in the sample);

- and robustness (whether the result is altered by small variations in operating conditions, such as temperature and the pH of reagents).

15.5 Bioburden specifications in the pharmacopoeias

The European, United States and Japanese pharmacopoeias have, for many years, all described procedures for bioburden determinations which, though similar in principle, have differed in detail. Harmonization procedures have now been agreed such that the main quality standards operative when this book is published should be those described in Table 15.4.

15.6 Identification of microorganisms

The terms identification and detection may sound much the same but they have different meanings and they are procedures undertaken in different circumstances. Detection tests, as described above, are routinely applied to raw materials and finished, manufactured medicines and they simply answer the question of whether a particular organism is present or absent. Identification of organisms, on the other hand, is a procedure usually undertaken together with environmental monitoring (Chapter 17). The regulators (FDA and MHRA) would expect that the major contaminating organisms in successive batches of nonsterile medicines should be identified in order both to assess their pathogenic potential and, in some cases, their likely source. Identification would also be required in the rare circumstances when a medicine showing signs of microbial growth and spoilage was returned to the manufacturer following a customer complaint.

Identification of bacteria is based upon several sources of information:

1. The appearance of the colonies growing on a Petri dish (e.g. shape, size, colour and surface markings).
2. The growth characteristics of the organism (for example, gaseous requirements – whether it grows aerobically, anaerobically or both; and its preferred growth temperature – if, for example, it were 25 °C the

Table 15.4 Quality criteria for nonsterile medicines.

Route of administration[a]	Maximum total aerobic microbial count CFU/g or ml	Maximum total yeast and mould count CFU/g or ml	Specified microorganisms absent in 1 g or 1 ml
Nonaqueous oral products	10^3	10^2	Absence of *E. coli*
Aqueous oral products	10^2	10^1	Absence of *E. coli*
Rectal products	10^3	10^2	
Oral mucosal, gingival, cutaneous, nasal and ear products	10^2	10^1	Absence of *Staph. aureus* and *Ps. aeruginosa*
Vaginal products	10^2	10^1	Absence of *Staph. aureus*, *Ps. aeruginosa* and *Candida albicans*
Inhalation products (excluding nebulized liquids)	10^2	10^1	Absence of *Staph. aureus*, *Ps. aeruginosa* and bile-tolerant Gram-negative bacteria

[a]Further criteria apply for transdermal patches, herbal medicines and natural products of animal, vegetable and mineral origin.

organism would be unlikely to be a pathogenic species because they usually prefer body temperature).

3. The microscopic appearance and staining reactions of the individual cells (for example, their shape, size, aggregation patterns, motility, spore formation and Gram stain result).

4. Biochemical tests that record the organism's ability to produce particular enzymes (for example, protease, urease and β-galactosidase), or utilize specific sugars as energy sources.

5. Immunological tests may occasionally be useful to confirm the identity of organisms like *Salmonella*, which are difficult to distinguish on the basis of their appearance and biochemical tests from much less hazardous ones like *Citrobacter* and *Edwardsiella* species.

6. Genetic analysis is becoming increasingly useful in the identification of fungi (see below) and can distinguish different strains of a bacterial species; this may be important, for example, when attempting to find out whether the very same organism has arisen in two different places or in different product batches.

In some cases it may be possible to identify a contaminant as far as genus level just by using the first three stages above – an aerobic, rod-shaped, spore-forming bacterium is almost certain to be a *Bacillus* species for example – and sometimes this level of categorization is sufficient; Gram-negative bacilli and Gram-positive cocci are other examples of nonspecific categorization. In most cases, however, it is necessary to identify the precise species, and this usually requires the fourth step: biochemical tests. These are often employed in the form of commercially available miniaturized test kits that enable multiple tests to be performed together. Protozoa hardly ever arise as contaminants, and viruses, though more likely, are very difficult to identify and are usually disregarded. This only leaves fungi as the other common contaminants requiring identification. Many yeasts can be identified in a similar manner to bacteria, and again there are test kits available, but the identification of moulds is a much more difficult and specialized undertaking; it is usually based upon their microscopical appearance, particularly of the spore-bearing structures and, increasingly, on genetic analysis using the polymerase chain reaction. If a pharmaceutical company or a hospital manufacturing unit required mould identification it is likely that an external specialist laboratory would be given the job.

Biochemical testing as a means of bacterial identification is not new. Traditionally it was done with test tubes and Petri dishes and it can still be performed that way, but when there are many isolates to identify such methods are time-consuming and require a great deal of media preparation, glassware, plasticware and incubator space. The use of commercially available test kits is, by far, the most common approach to bacterial identification and there are many available on the market although the Analytical Profile Index (API) products of bioMerieux are the most well established. Before using API kits it is

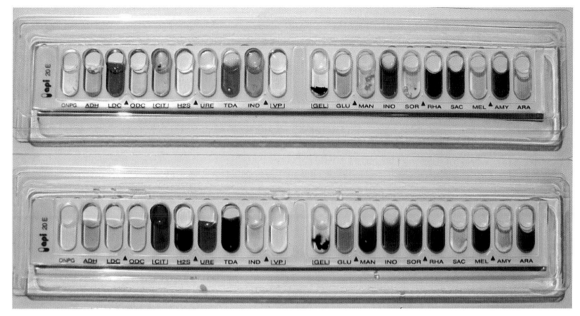

Figure 15.7 API Identification strips for *E. coli* (top) and *Proteus vulgaris* (bottom).

still usually necessary to Gram stain the organism, examine it microscopically and consider its likely source in order to select the most appropriate kit – of which there are many, for example kits for enteric bacteria, staphylococci, *Bacillus* species, anaerobic bacteria, nonfermenting Gram-negative rods, yeasts and so forth. The test organism has to be recently grown, standardized to a particular concentration and inoculated into cupules on the API gallery of tests (Figure 15.7). Following incubation, reagents are added where necessary, and the results (usually indicated by colour changes in the liquids) are recorded and converted into a seven-digit number. This is looked up either in a book or on a computer database to reveal the likely identity of the organism together with an indication of the reliability of the result.

Vitek is a more sophisticated system which operates on the same principles but is automated with respect to incubation, result recording and interpretation; it therefore eliminates some of the shortcomings of the API manual system where an operator's subjective interpretation of a colour change can significantly influence the result – several of the tests rely on distinguishing shades of yellow, orange and red. Vitek enables some bacteria to be identified within three hours and typically fewer than 2% are misidentified or not recognized at all; this permits the regular identification of bacterial isolates originating from many product samples and locations within a factory. There are several other manufacturers of similar systems and instruments having comparable specifications to API and Vitek.

Acknowledgement

Chapter title image: recording the number of bacterial colonies on a Petri dish. US National Archives and Records Administration #546274.

Chapter 16

Antiseptics, disinfectants and preservatives

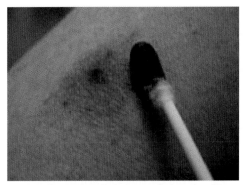

An iodine antiseptic being applied to the skin

KEY FACTS

- Chemical biocides are used as antiseptics, preservatives and disinfectants.
- They differ from antibiotics in being entirely chemically synthesized and also causing more general damage to cells.
- The rate and extent of activity is influenced by concentration, temperature, solubility, pH and other factors.
- Antimicrobial activity can be measured using qualitative or quantitative methods comprising broth dilution, agar diffusion and kill curve techniques.

Antimicrobial agents have traditionally tended to be divided into two groups: antibiotics and other chemical biocides. This distinction was primarily based on the fact that antibiotics were historically only of microbial origin while the chemical biocides were manufactured, synthetic products. In addition, antibiotics tended to have defined mechanisms of action whereas chemical biocides caused more general, nonspecific damage. In recent years these distinctions have become more blurred as antibiotics are often wholly chemically synthesized and chemical biocides have been found to have specific mechanisms of action.

However, we will continue with this distinction here as it is still of some value. Information on antibiotics is given in Chapter 10, so in this chapter we will deal just with the chemical biocides. These agents are employed as antiseptics, preservatives or disinfectants and are not used systemically for therapeutic purposes.

16.1 Definitions

16.1.1 Antiseptics

These are agents that have a broad spectrum of antimicrobial activity but are sufficiently nontoxic to allow them to be used on broken skin or mucosal surfaces.

16.1.2 Disinfectants

Again, these agents will possess broad antimicrobial activity but will have toxicity issues limiting their use

Essential Microbiology for Pharmacy and Pharmaceutical Science, First Edition. Geoffrey Hanlon and Norman Hodges.
© 2013 John Wiley & Sons, Ltd. Published 2013 by John Wiley & Sons, Ltd.

to inanimate surfaces such as worktops, floors, sinks and drains.

16.1.3 Preservatives

These are broad spectrum antimicrobial agents incorporated into pharmaceutical and other products to prevent the growth of contaminant microorganisms which might arise during use.

There is a common misconception that preservatives merely stop contaminant microorganisms from growing, but as we will see later the pharmacopoeial *Preservative Efficacy Test* requires that preserved formulations do actually bring about a *reduction* in the number of viable cells used to challenge the product.

Bactericidal, virucidal or fungicidal agents are those that kill bacteria, viruses and fungi respectively, whereas the terms **bacteriostatic, virustatic and fungistatic** refer to agents that merely arrest the growth of these microorganisms. In reality there is not a clear boundary between an agent which kills and one which arrests growth. These distinctions are often simply time and concentration dependent.

16.2 Classes of biocidal agents

Table 16.1 summarizes the characteristics of the many different classes of biocides and it is not the intention here to explore these in any greater detail. For further in-depth information on the nature of these agents you should refer to the more comprehensive texts given on the web site.

16.2.1 Principal target sites

Unlike antibiotics, which have fairly specific targets such as the synthesis of cell wall peptidoglycan, inhibition of protein synthesis and disruption of DNA function, chemical biocides tend to act is a more general manner. Their principal mechanisms of action are disruption of cell wall and membrane structure and function, intracellular coagulation and chemical modification of cellular proteins and nucleic acids. For this reason resistance to biocides tends to occur less readily than is found with antibiotics.

16.2.2 Factors influencing the activity of biocidal agents

A biocide interacting with a cellular target to bring about the death of that cell is a chemical reaction and so the rate and/or extent of that reaction is governed by the normal laws of chemistry. Hence, factors such as temperature, concentration, pH, solubility and so on will all influence antimicrobial activity.

16.2.2.1 Temperature

As with any chemical reaction, the speed of that reaction increases with an increase in temperature within certain limits. For antimicrobial agents the term used is *temperature coefficient* (Q_{10}), which describes the change in rate of kill for a 10 °C change in temperature. As an example, the phenols have a Q_{10} of around 4, which means that their rate of kill increases by a factor of 4 for every 10 °C rise in temperature. Clearly, there are limits to this because a temperature above about 40 °C may, itself, start to have detrimental effects on the cell.

16.2.2.2 Concentration

This is an extremely important factor which is frequently overlooked or misunderstood by people working with biocides. There is a natural assumption that doubling the concentration of a biocide will double the rate of kill, or conversely halving the concentration will reduce the activity by a factor of 2. While this rule may apply for simple chemical reactions the interaction of a biocide with cellular targets is much more complex. For biocides the term used is *concentration exponent* (η), which describes the change in rate of kill with a change in concentration. It is calculated using the following equation:

$$\eta = \frac{\log t_2 - \log t_1}{\log C_1 - \log C_2}$$

In this equation, C_1 and C_2 represent the concentrations of agent required to kill a standard inoculum in times t_1 and t_2. The concentration exponent η represents the slope of the line when log death time (t) is plotted against log concentration (C).

Table 16.2 gives some examples of concentration exponents of commonly used biocides. For those agents which have a concentration exponent of around 1, a doubling in concentration will bring about an increase in activity (kill) of 2^1, which is 2. However, for agents like the phenols, which have a concentration exponent of about 6, a doubling in concentration will bring about an increase in activity of 2^6, which is 64-fold.

The converse is that a halving in concentration will reduce their activity by 64-fold. Hence, these agents are

Table 16.1 Examples of different chemical biocides and their uses.

Biocide group	Examples	Spectrum of activity	Mode of action	Formulation issues	Commercial uses
Acridines	Aminacrine Acriflavine Proflavine	G + ve and G − ve cells Not sporicidal	Interfere with nucleic acid function	More effective at alkaline pH	Limited use in treatment of infected wounds
Alcohols	Ethanol Isopropanol Benzyl alcohol Bronopol Chlorbutanol Phenethyl alcohol Phenoxyethanol	G + ve and G − ve cells and fungi Not sporicidal and have low virucidal activity	Disrupt cell membranes	High concentration exponents Inactivated by organic matter Flammable	Widely used as antiseptics and preservatives
Aldehydes	Formaldehyde Glutaraldehyde Orthophthalaldehyde	Good activity against G + ve and G − ve cells, endospores, fungi and viruses	Cross link proteins by interacting with amino and other groups	Relatively high toxicity, particularly glutaraldehyde	Formaldehyde and orthophthalaldehyde used as disinfectants for medical equipment
Amidines	Propamidine Dibromopropamidine	Mainly G + ve cells and fungi Less active against G − ve cells and spore formers	Mode of action uncertain Inhibit oxygen uptake and induce amino acid leakage	Activity reduced by low pH and in blood and serum	Limited use in topical wound treatment
Biguanides	Chlorhexidine Alexidine Polyhexanide	Good activity against G + ve but less against G − ve cells and fungi Not sporicidal	Disrupt cell membranes	Incompatible with negatively charged excipients in formulation	Widely used as medical and veterinary antiseptics
Chelating agents	Ethylenediamine tetra-acetic acid	G − ve cells only	Increase permeability of cell wall of G − ve bacteria	Potentiates the effects of several antibacterial agents	Limited use as antibacterial agents Used to stabilize formulations

(continued)

Table 16.1 (Continued)

Biocide group	Examples	Spectrum of activity	Mode of action	Formulation issues	Commercial uses
Esters	Methyl, ethyl, butyl, propyl and benzyl parabens	Mainly G + ve bacteria and fungi Less active against G − ve cells	Not well understood Disrupt membrane transport processes; inhibit nucleic acid synthesis and inactivate key enzymes	Activity increases with alkyl chain length but solubility decreases Partition into oil phase of emulsions	Widely used as preservatives in pharmaceutical industry
Halogens	Chlorine Hypochlorites Iodine Iodophors	Broad antimicrobial spectrum Sporicidal	Cause enzyme and protein damage by interacting with amino and thiol groups	Can be irritant and staining	Used in skin disinfection and as general disinfectants
Isothiazolones	Range of commercial mixtures	Broad spectrum antibacterial, fungicidal	Inhibit active transport and glucose oxidation by binding to thiol groups on enzymes	Water soluble, pH stable and biodegradable	Mainly used as preservatives
Metals	Copper Mercury Silver Phenylmercuric nitrate (PMN) and Phenylmercuric acetate (PMA) Thiomersal	Phenylmercuric nitrate (PMN) active against G + ve and G − ve cells and fungi Not sporicidal	Silver binds with thiol groups on proteins and enzymes Interacts with bases on DNA	Toxicity problems with mercurials in particular. PMN incompatible with a number of common excipients Activity of silver depends on presence of Ag$^+$ ion	PMN and PMA limited use as preservatives Silver used as topical antiseptic and wound treatment
Organic acids	Benzoic acid Sorbic acid	Mainly active against fungi More limited activity against bacteria	Uncoupling agents Prevent uptake of substrates requiring proton motive force for transport	Activity highly pH dependent Only active at pH lower than 5	Used as preservatives particularly in the food industry
Peroxygens	Hydrogen peroxide Peracetic acid	Broad spectrum activity Sporicidal	Oxidation of functional groups on proteins	Hydrogen peroxide unstable	Used as antiseptics and disinfectants

(continued)

Table 16.1 (*Continued*)

Biocide group	Examples	Spectrum of activity	Mode of action	Formulation issues	Commercial uses
Phenols	Phenol Chlorocresol Chloroxylenol Triclosan	G+ve and G−ve cells. Slowly active against spores and acid-fast bacteria	Disrupt cell membranes Cause general cytoplasmic coagulation	High concentration exponents Some have limited solubility and can be adsorbed to polymers	Used as antiseptics, disinfectants and preservatives
Quaternary ammonium compounds	Benzalkonium chloride Benzethonium chloride Cetrimide Cetylpyridinium chloride	Broad spectrum antibacterials More active against G+ve than G−ve Some antiviral and antifungal activity Not sporicidal	Disrupt cell membranes	Incompatible with negatively charged excipients Benzalkonium chloride can cause sensitization	Widely used as antiseptics, disinfectants and preservatives
Quinolines	8-hydroxyquinoline Dequalinium chloride	Active against G+ve bacteria Less active against G−ve cells Some antifungal activity	Rapid uptake into cells Disrupt nucleic acid function	Some have low water solubility	Used as antiseptics and formulated in lozenges for throat infections
Anionic surfactants	Sodium lauryl sulphate	Weak antimicrobial properties	Disrupt cell membranes	Can interact with positively charged excipients in formulation	Limited use as antibacterial agents Used for detergent properties

Table 16.2 Concentration exponents of commonly used biocides.

Antimicrobial agent	Concentration exponent
Hydrogen peroxide	0.5
Mercurials	0.03–3.0
Bronopol	0.7
Iodine	0.9
Acridines	0.7–1.9
Quaternary ammonium compounds	0.8–2.5
Polymeric biguanides	1.5–1.6
Chlorhexidine	2
Parabens	2.5
Sorbic and benzoic acids	2.6–3.2
Phenols	4.0–9.9
Aliphatic alcohols	6.0–12.7

highly susceptible to losses in activity within a formulation due to adsorption, degradation and so forth. A lack of understanding of the concentration exponent can lead to the simple additive effect between two biocides being interpreted as synergy.

16.2.2.3   pH

The pH of a formulation will have a profound effect on the activity of some biocidal agents. Weak organic acids such as benzoic and sorbic acids will be highly ionized at alkaline pH values and in this form they cannot cross the biological membranes of the cell so their activity is poor. Only at low pH values will sufficient molecules be in the unionized form to allow good antimicrobial activity.

16.2.2.4   Solubility

This is an issue with classes of molecules having variable alkyl chain lengths such as the parabens. As the alkyl chain length increases from methyl paraben to butyl paraben the activity increases but the aqueous solubility decreases. As a consequence the parabens are normally used as mixtures containing, for example, methyl and propyl parabens together. Care must be taken when these systems are used in multiphase products such as emulsions where the longer chain length compounds may preferentially dissolve in the organic phase. Since the aqueous phase is the most likely component to be susceptible to contamination, that will be left relatively unprotected.

16.2.2.5   Interaction with excipients and packaging materials

Biocidal agents that are charged are very likely to adsorb onto oppositely charged excipients within a formulation. For example quaternary ammonium compounds carry a positive charge and will bind to negatively charged excipients such as alginates used as suspending or thickening agents. Similarly, adsorption to plastic or rubber components of the packaging materials or partition into the nonaqueous phase of an emulsion will result in a decrease in the concentration of biocide in the aqueous phase.

For each of the examples given above even a relatively small reduction in concentration of a biocide with a high concentration exponent could have severe consequences in terms of its ability to preserve the product. This will be discussed in greater detail in Chapter 17.

16.3 Measurement of antibacterial activity

Antimicrobial activity can be measured in a variety of ways each with their own advantages and disadvantages. While chemical assays can determine the concentration of an agent in solution, they cannot give information about the antimicrobial efficacy of the biocide and will not be considered further here. For information on antimicrobial efficacy it is necessary to evaluate the inhibitory effect of the formulation on live cultures.

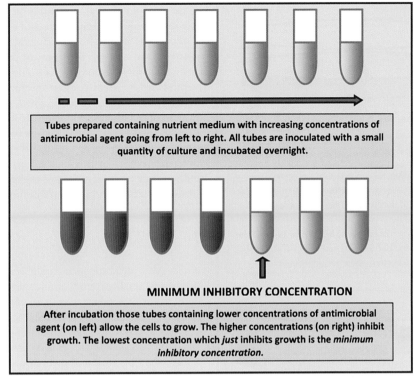

Tubes prepared containing nutrient medium with increasing concentrations of antimicrobial agent going from left to right. All tubes are inoculated with a small quantity of culture and incubated overnight.

MINIMUM INHIBITORY CONCENTRATION

After incubation those tubes containing lower concentrations of antimicrobial agent (on left) allow the cells to grow. The higher concentrations (on right) inhibit growth. The lowest concentration which *just* inhibits growth is the *minimum inhibitory concentration*.

Figure 16.1 Determination of the minimum inhibitory concentration (MIC) of a biocide.

16.3.1 Broth-dilution methods

If a bactericidal concentration of antimicrobial agent is diluted it will eventually reach a concentration at which it has no effect on a population of cells. Some of the cells may be killed but there will not be enough molecules of agent to damage the population as a whole. The end point of the dilution process can therefore be titrated to work out the lowest concentration of biocide which *just* inhibits the growth of a population of cells. This is termed the *minimum inhibitory concentration* or MIC (see Figure 16.1 and Figure 13.2). MIC values are often quoted as if they are constants but they are not. The values will vary markedly depending upon the incubation conditions, growth medium, strain of culture used, its growth history etc. In addition, the biocide dilution sequence occurs in steps, with sometimes a factor of 2 or even 10 between successive concentrations in the series. So, for example, if the MIC by experimentation is determined to be 1 μg/ml and the next concentration down in the series (but showing growth) is 0.5 μg/ml, all we can really say is that the MIC is between 0.5 and 1 μg/ml. This is because it could be 0.6 μg/ml but that concentration has not been tested.

16.3.2 Agar diffusion methods

This technique is commonly used to assay antibiotics and it is also used as a means of determining the sensitivity of clinical isolates to a particular antibiotic prior to treatment (Chapter 13). However, it is also used as a means of assessing the activity of chemical biocides. Figure 16.2 shows the basic principles of the process.

When a filter paper disc impregnated with biocide (or antibiotic) is placed upon an agar surface the biocide will dissolve in the water component of the agar and diffuse away from the disc. A concentration gradient will be established (the profile of which will change over time), being highest close to the disc and progressively less as we move towards the edge of the plate. While the biocide is diffusing, the bacteria within the agar will start to grow. Those bacteria close to the disc will experience high concentrations of biocide whereas those further away will initially not see any biocide molecules at all. As the concentration gradient is established there will effectively be an inhibitory front moving through the agar preventing growth of bacteria as it does so.

The formation of biocide diffusion zones is therefore a dynamic process involving diffusion of biocide and

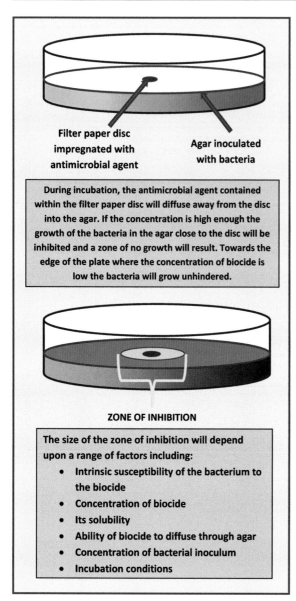

Filter paper disc impregnated with antimicrobial agent

Agar inoculated with bacteria

During incubation, the antimicrobial agent contained within the filter paper disc will diffuse away from the disc into the agar. If the concentration is high enough the growth of the bacteria in the agar close to the disc will be inhibited and a zone of no growth will result. Towards the edge of the plate where the concentration of biocide is low the bacteria will grow unhindered.

ZONE OF INHIBITION

The size of the zone of inhibition will depend upon a range of factors including:

- **Intrinsic susceptibility of the bacterium to the biocide**
- **Concentration of biocide**
- **Its solubility**
- **Ability of biocide to diffuse through agar**
- **Concentration of bacterial inoculum**
- **Incubation conditions**

Figure 16.2 Determination of biocide activity using zones of inhibition.

growth of bacteria. Eventually, the inhibitory front will reach a point in the agar where the numbers of bacteria have grown too high (*critical concentration*) to inhibit, and at that point the edge of the zone will be established. If the starting concentration of cells in the agar is higher, then it will take less time for this critical concentration to be achieved. Hence the inhibitory front will not have progressed so far and so the zone size will be smaller. Similarly, if the biocide concentration is higher, the inhibitory front will have reached further into the agar

by the time the critical concentration has been reached and so the zones of inhibition will be larger.

From this it follows that if everything else is kept constant, the size of the zone of inhibition will be dependent upon the biocide concentration and so this method can be used to assay the biocide. Figure 13.1 shows typical growth inhibition zones, albeit ones created by antibiotics rather than biocides. However, it should be borne in mind that agar diffusion data are often misinterpreted due to the erroneous belief that zones of the same size mean that two biocides are equally effective, and to a lack of appreciation of the effects of solubility, diffusion coefficients and concentration exponents.

16.3.3 Kill curves

Whilst MIC determinations and agar diffusion techniques give qualitative or semiquantitative information about the antimicrobial activity of a biocide they do not give any quantitative details on the rate at which the agent kills cells and how this might be influenced by environmental factors. To do this it is necessary to inoculate a biocide solution with a known concentration of viable cells and then take samples at intervals of time to determine the number of surviving cells at each time point (see Figure, 16.3 and 18.1).

Of particular importance here, and a point often overlooked, is that when a sample is taken the antimicrobial agent must be neutralized so that when placed upon agar to evaluate survival there is no residual activity to inhibit cell growth. The pharmacopoeias give a range of options for neutralization depending upon the nature of the biocide; these may include dilution (for those with high concentration exponents), specific chemical inactivation or general neutralization with a combination of lecithin and Tween 80 (Table 15.1).

16.4 Preservative efficacy testing

The rationale behind the use of preservatives in pharmaceutical and cosmetic products is described in more detail in Chapter 17. Here we will just give information on the pharmacopoeial testing procedure for the final products. Due to the complex nature of many pharmaceutical and cosmetic products it is not sufficient just to perform a chemical assay to determine the concentration of antimicrobial agent in the formulation. As we have previously mentioned, factors such as pH,

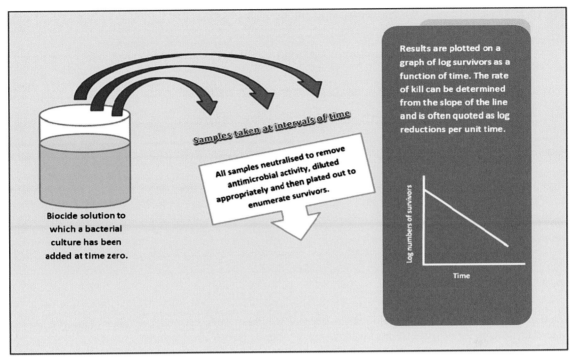

Figure 16.3 Determination of biocide activity using kill curves.

multiphase systems, presence of suspended solids and potential adsorption onto packaging materials can all influence the biological activity of the preservative. This must be tested when the product is first manufactured and also during storage to show there is no deterioration of preservative action over time.

Preservative efficacy tests are described in detail in the various pharmacopoeias and it is important for the operators to adhere strictly to the requirements given. Only the broad concepts will be outlined here and the reader is referred to the pharmacopoeias for more information.

The preservative efficacy test is essentially run along the same lines as the kill-curve methodology explained above. The product is tested in its final container and different samples are inoculated with a range of different cultures. The final concentration of cells in the test sample and the volume ratio of product to inoculum are all described in detail. The organisms used (including specified strain numbers) are generally the same for the main pharmacopoeias and include:

Pseudomonas aeruginosa	*Candida albicans*
Staphylococcus aureus	*Aspergillus braziliensis*
Escherichia coli	*Zygosaccharomyces rouxii*

After inoculation of the products in their final containers they are stored under specified conditions for 28 days. During this period samples are removed at intervals of time and neutralized before enumerating the survivors using plate counts (Chapter 15). Bacteria are incubated at 30–35 °C for 18–24 hours; *Candida* and *Zygosaccharomyces* at 20–25 °C for 48 hours and *Aspergillus* at 20–25 °C for one week. The pharmacopoeias define the performance criteria required for different product types and for each of the various microorganisms which have challenged the product.

The requirements given in the EP are shown in Table 16.3 and it can be seen that, although the term 'preservative' implies that this will simply prevent growth, the system must be able to kill those challenge organisms. Note that there are two sets of acceptance criteria for parenteral/ophthalmic and topical products. The *A* criteria are more demanding and are the ones generally applied. However, there may be some circumstances when the less stringent *B* criteria may be acceptable, for instance if there is a risk of toxicity if the concentration of preservative is too high.

From the table it can be seen that in order to pass the preservative efficacy test, a topical product, for example, would have to bring about a 2-log reduction in viable count of all the challenge bacteria within 48 hours; a 3-log

Table 16.3 EP performance criteria for preservative efficacy tests.

Product type	Challenge organism	Acceptance criteria	Log reductions specified					
			6 h	24 h	48 h	7 days	14 days	28 days
Parenteral/ophthalmic	Bacteria	A	2	3				NR
		B		1		3		NI
	Fungi	A				2		NI
		B					1	NI
Topical, nasal, ear and inhaled products	Bacteria	A			2	3		NI
		B					3	NI
	Fungi	A					2	NI
		B					1	NI
Oral and rectal products	Bacteria						3	NI
	Fungi						1	NI

NI = no increase; NR = no recovery.

reduction at 7 days and the count should remain at that low level for 28 days. Similarly the counts for all the fungi should be reduced by 2-logs at 14 days and be shown not to have increased from that value when tested at 28 days.

16.5 Disinfectant testing

Most people will know what they mean by the term disinfectant but in fact there is no universally accepted definition. Clearly, there is a need for a disinfectant to exert a much greater antimicrobial effect than a preservative and this generally implies the reduction of microbial load to a level where it no longer poses any threat. Note that it does not imply sterility. As disinfectants are intended for use within diverse environments there is a requirement that they possess bactericidal, sporicidal, fungicidal and virucidal properties.

Disinfectant testing is by no means straightforward with numerous published standards and it is beyond the remit of this book to give details of each and every protocol. As with the preservative efficacy test, the basic approach is to add microorganisms to a disinfectant and remove samples at intervals of time, neutralize the biocide and assess the survivors. For simplicity we will divide the protocols into semiquantitative and quantitative tests.

16.5.1 Semiquantitative tests

In these tests the samples removed at intervals after initial challenge are not enumerated but simply added to broth,

incubated and observed for any signs of growth. The tests can be further divided into:

- Suspension tests
  - This is the simplest test where the microorganisms are added to disinfectant diluted with water and the removed samples inoculated into broth containing a neutralizer. An example of this type of test is the Rideal Walker test.
- Capacity tests
  - This is an improvement on the simple suspension test; it is designed to mimic repeated challenge of the product such as might be seen with a dirty mop being constantly placed into a bucket of disinfectant. The disinfectant is subjected to multiple challenges of test microorganisms (mixed with yeast suspension to simulate organic soil) and samples removed for testing as before. An example of this type of test is the Kelsey–Sykes test.
- Surface tests
  - These tests are designed to evaluate the effect of disinfectant on microorganisms dried onto surfaces. The specified microorganisms are dried onto either stainless steel or borosilicate glass cylinders. The cylinders are then placed into a solution of disinfectant normally in the presence of 5% serum (again to simulate soiling) for a given period of time. The cylinders are then removed and placed into broth containing a neutralizer and incubated to determine if there are any survivors.

16.5.2 Quantitative tests

These are fully quantitative tests where the samples are enumerated to find out the number of surviving cells. From these tests it is possible to determine the rate of kill for a particular disinfectant and one generally accepted criterion is the ability of a disinfectant to achieve a 5-log kill within 5 minutes. The tests can be divided into two phases:

- Phase 1 tests
  - These are simple suspension tests where the disinfectant is diluted in distilled water with no added soil. They are designed to just check that the product has some antimicrobial activity.
- Phase 2 tests
  - These are designed to mimic in-use conditions and are further divided into Step 1 and Step 2 tests.

- Step 1 tests are similar to Phase 1 tests but use a wider range of challenge organisms, contain organic soil and the disinfectant is diluted in hard water.
- Step 2 tests are simulated in-use tests and include surface disinfection testing as outlined above but, of course, with a quantitative rather than qualitative approach.

The authorities are striving to develop full in-use field tests but these are proving much more difficult to standardize.

Acknowledgement

Chapter title image: http://commons.wikimedia.org/wiki/File:ExAntiseptic.jpg; PD-USGov-Interior-USGS-Minerals.

Chapter 17

The manufacture of medicines: product contamination and preservation

A medicines' filling line (largely enclosed to protect against contamination)

KEY FACTS

- The numbers and types of microorganisms present in a nonsterile medicine are controlled both to avoid the product becoming an infection hazard and to minimize degradation of the active drug or the excipients.
- Sterile medicines can be made in two ways: either by terminal sterilization, or by aseptic manufacture which involves the mixing of individually sterilized ingredients under sterile conditions.
- Even in terminally sterilized products it is important to minimize contamination during manufacture in order to avoid high endotoxin levels from dead bacteria.
- Monitoring the levels of organisms in the manufacturing environment is standard practice in the pharmaceutical industry, so data are routinely recorded for levels of bacteria and fungi in the atmosphere, and on working surfaces, equipment, personnel and their protective clothing, as well as in water and raw materials.
- Medicines and medical devices are made in 'clean rooms' where the levels of microorganisms in the atmosphere are carefully controlled.
- The manufacturer of a medicinal product is responsible for the quality of that product throughout its user life.

Essential Microbiology for Pharmacy and Pharmaceutical Science, First Edition. Geoffrey Hanlon and Norman Hodges.
© 2013 John Wiley & Sons, Ltd. Published 2013 by John Wiley & Sons, Ltd.

- Multiple-use products must be formulated to prevent the growth of microorganisms that arise as contaminants during use. This is usually achieved by the addition of chemical preservatives, although other methods will be discussed.
- Care must be taken to ensure that the chemical preservative does not interact with the components of the formulation.
- Knowledge of the concentration exponent of the biocide is important to be able to appreciate the consequences of any loss of preservative.

17.1 Microbiological standards for medicines

It is obvious that there must be standards for the microbiological quality of medicines in order for them to satisfy the fundamental requirements of the regulatory authorities: quality, safety and efficacy. It is obvious, too, that in addition to these criteria the product must be appealing to the consumer; a product having a pleasant taste, smell and appearance is more likely to be commercially successful than one that hasn't. The bioburden of a manufactured medicine must therefore be controlled because:

- the presence of pathogenic organisms in the medicine would make it an infection hazard;
- contaminating microorganisms may degrade the active ingredient of the medicine and thereby effectively reduce the dose that the patient receives;
- contaminants may alter the physical stability of the medicine (by, for example, breaking down emulsifying agents and causing creams or lotions to separate into oil and water layers, or by breaking down thickening agents which control the flow rate of liquid medicines from a bottle);
- a medicine that is obviously contaminated with microorganisms would be unacceptable to a consumer in just the same way that mouldy food is unacceptable.

From a microbiological perspective medicines can be divided into two categories: sterile and nonsterile. The main sterile ones – those containing no living organisms at all – are injections and those applied to the eye. Other less commonly used products like irrigation solutions and those introduced into the ear may also be sterile. Virtually all other products are nonsterile but, of course, that does not mean there is no control on the number or types of organisms that may be present; the microbiological quality standards that apply are those described in Chapter 15 on bioburdens.

17.2 Methods of making sterile products

A sterile product is made in one of two ways:

- terminal sterilization: the product is made completely, packed into its final container and then sterilized;
- aseptic manufacture: the various ingredients of the product may be individually sterilized (usually by passing solutions through bacteriaproof filters) and then mixed together using sterile equipment under conditions that do not allow the entry of microorganisms.

Terminal sterilization is the preferred strategy, primarily because it is safer and more reliable, although it may also be more convenient and cheaper too; despite this, the majority of sterile products are made by aseptic manufacture.

Steam is by far the most commonly used method of terminal sterilization, but many sterile medicines contain one or more heat-sensitive ingredients so aseptic manufacture would be the most likely option. Sometimes the product as a whole might be heat sensitive: an ophthalmic cream, for example, must be sterile, but if it were heated it would cease to be a cream, so that too would probably be manufactured aseptically. The process requires extraordinarily high standards of hygiene and rigorous adherence to operating protocols in order to ensure that every batch is safe. That is not to say that a terminally sterilized product can be made in such a way that significant microbial contamination might be introduced during manufacture in the belief that the organisms would all be killed at the end anyway. Quite apart from the fact that dead bacteria are still pyrogenic (cause a rise in body temperature) so an injection would be at risk of failing the pharmacopoeial endotoxin test, the regulatory authorities do not condone a strategy whereby poor manufacturing practices are covered up by terminal sterilization or the use of preservative chemicals.

Not surprisingly, the standards that apply to non-sterile medicines – the permitted levels of microorganisms in the air of the manufacturing environment for example – are less exacting than those that apply to sterile products but, nevertheless, a considerable amount of attention is given to cleaning, disinfection and hygiene in order to satisfy the requirements of the pharmacopoeias and the regulatory authorities. All medicines, even nonsterile ones, are generally manufactured under much stricter codes of practice and to more exacting microbiological standards for the finished product than those applicable to food manufacturers for example.

17.3 Strategies to assure appropriate standards

There is not, of course, one single action that can ensure that high microbiological standards are achieved; attention must be given to all aspects of the manufacturing process and this is, itself, a subject on which books are written, so it is not possible here to discuss in detail all of the policies that might be implemented. However, some of the more important aspects include:

- The use of high quality raw materials. A pharmaceutical manufacturer is likely to set a specification for a raw material that covers not just the minimum assay value and limits for chemical contaminants but the minimum achievable levels of microbial contaminants too.
- Environmental monitoring. The manufacturer will routinely monitor the levels of contamination in the manufacturing environment. This will include sampling of air, equipment, work surfaces, personnel protective clothing and the water used in manufacturing and cleaning; results will be recorded, charted, subjected to statistical trend analysis and made available to inspectors.
- Cleaning and sanitization. Procedures for cleaning and disinfection will be precisely described, together with validation strategies to confirm that the procedures do, in fact, effectively control levels of microbial contamination.
- Avoidance of conditions suitable for microbial growth. Some degree of microbial contamination of raw materials is usually inevitable, but reproduction of those organisms is prevented by avoiding, where possible, storage or maintenance of aqueous solutions at neutral pH in warm conditions – particularly if the solutions contain proteins or carbohydrates. Water to be used for

manufacturing is often maintained at a temperature of $80\,°C$, and is passed through the pipes of a distribution system at a flow rate of 1–2 m/s to avoid the formation of bacterial biofilms.

- Increasing use of policies designed to identify and protect stages in the manufacturing process where problems might arise. For example, Hazard Analysis of Critical Control Points (HACCP) would identify the final stage in which an aseptically manufactured medicine is filled into open vials or ampoules as one of the most vulnerable points because the liquid is no longer enclosed in a pipe but exposed to possible contamination from the atmosphere.
- Appropriate use of chemical preservatives in medicines that are vulnerable to microbial spoilage.
- Design of containers to avoid in-use product contamination. The intention should not be merely to avoid contamination during the manufacturing process itself but to restrict it whist the product is used by the patient or consumer – hence the increasing use of, for example:
  - cream and ointment tubes rather than the wide-mouthed jars that are vulnerable to contamination from the patient's fingers;
  - single-dose eye drops (to eliminate contamination from infected eyes of the droppers often used in multidose eye drops); and
  - individual dose packing – blister packs – for tablets, capsules and even injections in disposable syringes.
- Accurate and comprehensive record keeping enabling the history of a batch to be reviewed and analysed in detail if a problem should arise.

17.4 Sources of microbial contamination, and environmental monitoring

Microorganisms can arise in a manufacturing area from the atmosphere, equipment and work surfaces, personnel, water and from raw materials and their packaging. It is important to recognize that whilst each of these represents a potential *source* of contamination for a manufactured medicine, two of them, air and water, assume particular importance because they are also vectors which facilitate movement of organisms from one place to another, and water is yet more important not only because of its widespread use as a raw material and cleaning agent but also because it serves as a medium in which organisms can actually grow to high concentrations – up to 10^5 CFU/ml or more in some cases.

17.4.1 *The atmosphere*

Microorganisms in the air are usually attached to dust particles which, in a pharmaceutical factory, usually consist largely of skin flakes shed from the manufacturing personnel (see below). Bacteria, protozoa and viruses do not, therefore, normally exist in the air as individual cells or clumps of cells (or particles in the case of viruses). Fungal spores, however, are often released into the air from the fruiting body of a mould, so they are the exception in that they are not necessarily attached to dust, and air filters must be designed to trap such small particles. Microorganisms do not grow – reproduce – in air because it is too dry. Indeed, many organisms die on prolonged suspension in the atmosphere as a result of drying, oxygen toxicity (in the case of anaerobes) or even photosensitivity. Generally it is Gram-positive bacteria and spore-forming organisms (of both bacteria and fungi) that survive drying best; Gram-negative bacteria normally arise in aqueous environments and are much more sensitive to desiccation.

Because most of the dust comprises human skin scales, it is usually the case that the concentration of microorganisms in the air of a pharmaceutical 'clean room' (described below) is influenced by the number of operators and the extent to which they move around. The more movement there is, the greater the number of skin scales shed, and this has been estimated to vary between 100 000 particles per minute for a motionless person to 10 million per minute for a vigorously active one. Not all of these particles will necessarily carry viable microorganisms however, and suitable protective clothing can substantially restrict release of the particles into the air. Operators wear gowns (Figure 17.6) that cover as much skin surface as possible together with disposable gloves and coverings for hair and, where necessary, beards. Despite filtration, clean-room air is not sterile; there is usually a low level of microbial contamination, but the high efficiency particulate air (HEPA) filters commonly used are capable of removing over 99.99% of 0.3 μm diameter particles.

There are two approaches to the monitoring of levels of microorganisms in the atmosphere. Passive monitoring uses so-called 'settle plates' which are simply Petri dishes exposed to the atmosphere in prescribed locations for a fixed time, typically four hours, so that organisms can settle under gravity onto the moist surface of the agar and develop into visible, countable colonies after incubation. Usually two different culture media are employed: one suitable for the growth of bacteria and one for fungi (tryptone soya agar and Sabouraud

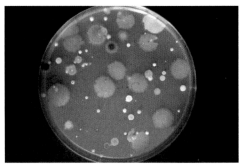

Figure 17.1 A typical settle plate showing the number and variety of microorganisms that may be deposited during a one-hour exposure in a university laboratory.

dextrose agar respectively are commonly used). The number of colonies obviously depends upon the concentration of organisms in the atmosphere where the plate was exposed, but the volume of air passing over the plate will influence the count too, so locations where there is naturally a high degree of air turbulence, by doors for example, might be expected to give higher counts. The number of colonies might be very high in a busy room with no air filtration, for example the university laboratory in which the plate in Figure 17.1 was exposed, but pharmaceutical clean rooms have a much lower level of atmospheric contamination so in a sterile manufacturing area there would normally be no colonies arising at all following a four-hour exposure.

Active air-sampling measures the concentration of organisms in the atmosphere in terms of numbers per litre, so most of the commercially available instruments have a pump that passes air through a slit and causes the suspended particles to impinge onto the surface of the agar in an open Petri dish or plastic strip. The crucial advantage of active sampling is that the result should be more reproducible and reliable because a known volume of air is sampled, but the instruments commercially available, which typically number 8–10 in the UK, are often relatively expensive, bulky and difficult to transport, and may be difficult to fumigate or disinfect. Those such as the Biotest sampler (Figure 17.2) which does not use standard 9 cm Petri dishes may also be relatively expensive to operate.

Regardless of whether settle plates or active air sampling are used, the colony count is normally plotted on a graph so that trends can be recognized, and limits would be adopted both to alert, and require action from, the staff responsible if the values increased abnormally.

Figure 17.2 Biotest air sampler and (right, and not to scale) the agar strip after incubation on which bacterial colonies have grown. Before sampling, the strip is fed into the circular housing at the top of the instrument; this contains an impeller to suck in air which is blown against the moist agar surface of the strip. Dust and attached bacteria are caused to 'stick' to the agar and grow into visible colonies during incubation.

17.4.2 Solid surfaces

Microorganisms may be expected on most solid surfaces in a manufacturing unit, so it is common to monitor contamination on walls, floors, bench tops and equipment. Much of the contamination arises from dust, so again Gram-positive bacteria and spore-forming organisms would be the most prevalent types; the number encountered would depend on the factors mentioned above and, in addition, the rugosity (roughness) of the surface, its charge and chemical properties. The surfaces of many metal, glass, plastic and other materials appear to the naked eye to be of similar smoothness, but when viewed under the microscope substantial differences are seen in terms of ridges, folds or contours and these may significantly alter the surface area to which dust and microorganisms can attach; obviously fabrics would represent an extreme example. Microbial cells normally have a negatively charged surface, but the charge on the dust particles to which the organisms are attached is of greater importance in determining whether there is a charge attraction or repulsion from a solid surface. Some materials may possess intrinsic antimicrobial activity or, in the case of plastics, even have antimicrobial agents deliberately incorporated in them.

Again, there are two methods available for monitoring surface contamination: swabs and contact plates (also known as replicate organism detection and counting – Rodac – plates). Sterile swabs are individually packed and simply comprise a stick, usually of wood, at one end of which is a fabric tip which is wetted with a suitable diluent then rubbed over a defined area of the surface to be examined. The swab is normally handled using sterile gloves so that no contamination from the operator's hands is transferred to the stick. After sampling, the whole swab is returned to the container of sterile diluent and thoroughly shaken to transfer into the liquid any organisms that have been collected from the sampled surface; viable counts (as described in Chapter 15) are then performed on the liquid. This method of sampling is convenient for irregular surfaces for which contact plates are unsuitable (see below) and is commonly employed for sampling curved surfaces of room coving or manufacturing equipment. It suffers from the disadvantage that it is not possible to be sure that all the organisms collected on cotton fabric swabs are washed off into the diluent; this may be overcome by the use of alginate swabs which dissolve completely in some diluents. It is important that the diluent selected should support the survival of any organisms on the swab during its transport it to the laboratory but should not contain nutrients that permit their growth and multiplication in numbers during that time. Isotonic phosphate buffered saline having a pH of 7.3 is a suitable diluent for most organisms.

Contact plates (Figure 17.3) are small Petri dishes the bases of which are filled completely with molten agar such that the surface of the medium, when set to a gel, is convex and extends above the plate (surface tension

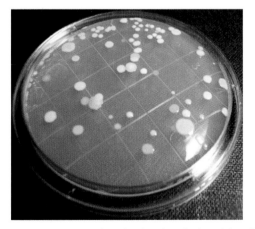

Figure 17.3 A contact plate showing the colonies arising after the plate had been used to sample a bench top. Note this plate is approximately half the diameter of the Petri dish shown in Figure 17.1.

enables the dish to be slightly overfilled without it over-flowing). Once set, the convex surface of the agar is pressed against the surface to be monitored and micro-organisms are transferred to the agar; the plates are then incubated and the colonies counted. The amount of pressure applied to the base when pressing the plate onto the surface will influence the proportion of orga-nisms that are transferred to the agar. Contact plates are much easier to use than swabs but are only suitable for flat surfaces so they are commonly employed for sam-pling the contamination on floors, walls and equipment.

17.4.3 Personnel

In addition to shedding skin scales into the air, operators in a manufacturing unit can introduce contamination into the product directly from their skin or from their nose in exhaled air. Skin contaminants are commonly micrococci, staphylococci, corynebacteria (also called diphtheroids) and propionibacteria (including the organisms associated with acne). Corynebacteria are quite variable in terms of the cultivation conditions they require (some are anae-robes), and as a consequence their numbers on the human skin tend to be underestimated. Staphylococci may be present both on the skin and in the nose of healthy individuals. The proportion of humans carrying *Staph. aureus* in the nostrils is often quoted in the range 25–40%, and its presence as a contaminant of manufactured medi-cines is usually regarded as a symptom of poor factory hygiene because operating personnel are the most likely source. Despite this, it is not common for manufacturing personnel to be asked to provide nasal swab samples because their noses and mouths are usually covered with a face mask anyway. Consequently, monitoring of contamination arising from operators is usually restricted to the external surfaces of masks, coats and gloves.

Organisms on the skin may be transferred directly onto the surface of agar in Petri dishes, so again contact plates may be employed but, more commonly, stan-dard-sized (9 cm) plates are used to estimate the degree of contamination from the fingers using the procedure known as 'finger dabs' whereby the finger tips are rolled across the agar surface in the same way that fingerprints are taken in police investigations. Figure 17.4 shows a finger dab plate, but it should be emphasized that this was prepared by direct skin sampling from unwashed hands simply to illustrate the relatively high numbers of organisms on hands that had not been washed or disinfected. Manufacturing personnel would often wear sterile gloves and the number of colonies arising should be very much smaller.

17.4.4 Water

The microbiological quality of raw materials is obviously likely to influence that of the final manufactured product so specifications for raw materials are likely to put limits on the maximum permissible concentrations of orga-nisms in particular categories (for example, total viable bacteria, yeast and moulds, *Enterobacteriaceae*) as well as specifying which objectionable organisms should not be present. Water, however, is sufficiently important to be worthy of special mention here, not only because it is the most commonly used raw material but also because it is used extensively for cleaning, so if water with a high level of microorganisms is used for rinsing detergents from work surfaces and manufacturing equipment that might negate the sanitizing effects of any disinfectants that may have been used earlier.

The organisms most commonly found in water are Gram-negative bacteria; Gram-positive species and fungi are usually present in much lower concentrations. Mains water contains chlorine which severely limits the num-bers of viable bacteria, so the total viable count is typically less than 100 CFU/ml in fresh tap water and often it is much lower even than this, although values vary depend-ing upon temperature and rainfall patterns. Once the chlorine has been removed by heating or deionization the water becomes vulnerable to bacterial growth because it is still likely to contain sufficient proteins and other nutrients to support populations of 10^5 CFU/ml or more. Stored water therefore is likely to have a higher count than water that has been recently purified. One problem that arises in manufacturing units is the growth

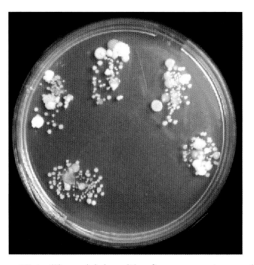

Figure 17.4 'Finger dabs' resulting from an operator pressing unwashed hands onto the surface of a Petri dish.

of bacterial biofilms on the inner surfaces of water pipes. Such biofilms are difficult to remove and may, therefore, represent a persistent source of contamination, so it is good practice to install water piping with sufficient access points for hypochlorite or other disinfectants to be introduced to kill the biofilm. Ensuring an adequate flow rate of water within the pipe may also restrict biofilm formation.

Pharmacopoeial specifications for purified water limit the bacterial concentration to 100 CFU/ml, and because it is this grade of water that is most commonly used as a raw material and for rinsing, bioburden determinations should be expected to give low results; the most suitable counting method, therefore, is membrane filtration (Chapter 15), although instrumental methods of counting that combine laser scanning of cells on filter membranes and epifluorescence are becoming more common. When using traditional plate counting methods the number of bacterial colonies that develop from a water sample will depend upon the culture media used, and although it is perhaps counterintuitive, the R2A agar medium recommended in the pharmacopoeias is nutritionally less rich than the tryptone soya agar which was formerly used but nevertheless usually permits a much larger number of bacteria to develop into visible colonies.

17.4.5 Raw materials used for manufacturing medicines

The subject of bioburdens in raw materials is covered in Chapter 15.

17.5 Clean-room design and operation

Medicines and medical devices are manufactured in a 'clean room', which is simply an environment with a controlled low level of atmospheric contamination (dust, microorganisms or chemical vapours). Air in the room is recirculated through HEPA filters which are often located in the ceiling and are designed to achieve a vertical laminar (unidirectional) air flow throughout the room (Figure 17.5).

Clean rooms are classified according to their air quality (see Table 17.1) and it is common for rooms dedicated to the most critical manufacturing steps, for example the filling of a sterile liquid into ampoules, only to be accessible to operators who have dressed in suitable protective clothing ('gowned-up') in outer rooms of

lower air quality and enter the critical areas through airlocks. Clean rooms, therefore, are often designed in suites with different zones permitting different activities. Steps are taken to avoid introducing dust and microbial contamination, not only by requiring the operators to wear gloves, gowns, overshoes, hair- and face-masks (all manufactured from fabric that does not shed fibres), but also by sterilizing or disinfecting items of equipment, containers, liquids and raw materials that enter the room. Even disinfectant solutions are diluted in sterile water and made up fresh on a regular basis to avoid them acting as sources or reservoirs of contamination with organisms resistant to the disinfectant, and the mops and buckets used for cleaning are specially designed and dedicated for the purpose. Furniture is kept to a minimum since that too might be a source of contamination and both the furniture and the room itself is designed for ease of cleaning. Surfaces are smooth, nonabsorbent, free of cracks and ledges, and made of material that could be disinfected (or even fumigated if necessary); stainless steel is the preferred construction material. Clean rooms normally operate at a slight positive pressure relative to the surroundings so that potentially contaminated air does not enter the clean room when the door is opened. Not only the pressure, but also the temperature and humidity of the air are controlled.

Clean rooms are expensive to build and need to be operated according to very strict rules in order to function well. In recent years there has been a trend towards the use of isolators as alternatives to clean rooms, not

Figure 17.5 Vertical laminar airflow in a clean room. *Source:* http://commons.wikimedia.org/wiki/File:Laminar_Flow_Reinraum.png.

Figure 17.6   Three isolators used for filling cytotoxic injections. Pierre Fabre cytotoxic injectable filling line. Reproduced with kind permission from Pierre Fabre Médicament Production/Aquitaine Pharm International (API) (www.aquitaine-pharm.com).

Table 17.1   Clean room classifications described in the European Orange Guide.

Grade	At rest		In operation	
	Maximum permitted number of particles per m^3 equal to or above			
	0.5 μm	5 μm	0.5 μm	5 μm
A	3500	1	3500	1
B	3500	1	350 000	2000
C	350 000	2000	3 500 000	20 000
D	3 500 000	20 000	Not defined	Not defined

only for critical manufacturing steps but also for selected quality control functions like sterility testing (see Chapter 19). Isolators provide a physical barrier between the operator and the work process, and this is the feature that distinguishes them from clean rooms. The barrier may be in the form of a flexible, transparent plastic enclosure which is supported on a metal frame, or it is a rigid, self-supporting structure with glass or plastic windows (Figure 17.6). In either case, the isolator is sealable and capable of being gassed, usually with hydrogen peroxide, to sterilize the inside. Materials entering the isolator are first sterilized or disinfected then passed through a hatch or port; the operator manipulates them via gloves sealed into the wall.

The classification of clean rooms is based upon the concentration of particles in the air, and there are several particle counting instruments on the market by which this can be measured. The situation is confused by the fact that there are different classification schemes: the International Standard ISO146441-1 and the British Standard 5295 specify particle concentrations as a number per cubic metre, whilst the officially cancelled but still widely quoted US Federal Standard 209E specifies numbers per cubic foot. In Europe, the clean-room classification of the *Rules and Guidance for Pharmaceutical Manufacturers and Distributors 2007* – the 'Orange Guide' must be used (Table 17.1). The specifications vary according to whether the room is unoccupied or in use, because higher levels of contamination might be expected where operators are present. A class A room would be used for the most critical manufacturing steps for a sterile product whilst rooms of lower classifications would be appropriate for nonsterile medicines. To put

the numbers specified in the table into perspective, 'ordinary' air in an urban environment would be expected to have at least 10 × the number of 0.5 μm particles as that in a class D clean room.

The concentration of particles in the air is not the only Orange Guide specification for the rooms described in Table 17.1; there are also maximum limits for the number of colonies arising on both settle plates and contact plates, and again there are different specifications depending upon the occupancy status of the room.

17.6 The protection of pharmaceutical products from microbial contamination

It is the responsibility of those producing medicines to ensure that they conform to good manufacturing practice (GMP) requirements for quality, safety and efficacy. However, in addition medicines should be elegant; they must be acceptable to patients and/or consumers and the quality must be maintained during their period of use. In other words the manufacturer is not just responsible for the product up to the time it leaves the factory gates but for the whole of the product shelf life.

Most of the medicinal products used clinically are nonsterile, which means that they may contain microorganisms and may be exposed to microorganisms during storage and use. While this is not a problem for many medicines, particularly those single-use products such as tablets, it does present difficulties for multiple-use products.

Table 17.2 Examples of different types of pharmaceutical products. The products highlighted in red are those that may be in need of a chemical preservative within the formulation.

Route of administration	Sterile/nonsterile	Mode of use	Examples
Parenteral	Sterile	Single use	Vials, injections and infusions
		Multiple use	Insulin, some vaccines
Ophthalmic	Sterile	Single use	Minims
		Multiple use	Most eye drops, ointments etc.
Urinary	Sterile	Single use	Bladder irrigations
Oral	Nonsterile	Single use	Tablets/capsules
		Multiple use	Liquids (solutions, syrups, suspensions, emulsions etc.)
Topical	Mostly nonsterile	Multiple use	Creams, ointments, lotions, gels, pastes, dusting powders
Respiratory	Nonsterile	Single use	Dry powders
		Multiple use	Liquid inhalers
Rectal	Nonsterile	Single use	Suppositories, enemas

Table 17.2 gives examples of the range of medicinal products available and highlights those sterile and nonsterile products, which, during their period of use, may be exposed to microbial contamination. A number of factors are associated with the risk to the product during its use and these are indicated below:

- Ease of entry into product or container.
  - Is it better to put a cream into a tube with a narrow neck rather than a wide mouth jar to minimize microbial exposure?
  - Should tablets be sealed in a blister pack rather than loose in a bottle?
- Type and magnitude of contaminant bioburden.
  - Natural products often have higher initial bioburden than synthetic products.
- Mode of intended use (potential for misuse).
  - If patients cannot easily access their medicine (child-proof lids or blister packs) will they be removed and left open to the atmosphere?
  - Elderly patients may have difficulty pouring oral solutions into a 5 ml spoon and hence drink directly from the bottle.
  - Many patients have problems self-administering eye drops and eye ointments leading to inappropriate use and potential for increased microbial exposure.
- How long is the product likely to be used/stored?
  - The manufacturer may stipulate storage conditions and period of use but has no control over how the

medicine is actually treated; this must be factored into the design of the product.

17.6.1 The consequences of microbial contamination

If a product is poorly designed or manufactured then it is liable to become heavily contaminated with microorganisms. The potential consequences of this are listed below:

- loss of aesthetic appeal;
- product deterioration;
- degradation of active ingredients;
- noncompliance;
- product recall;
- adverse publicity;
- harm to patient/consumer;
- litigation.

It is worth pointing out that product spoilage and even patients becoming infected from their medicines is not simply a theoretical possibility; there are many examples in the past where this has been a reality. While most of the items in the list above are self-explanatory we will highlight some of these issues here.

Loss of aesthetic appeal may take the form of obvious visible growth such as moulds on the surface of creams, changes in the turbidity of solutions and discolouration due to the production of microbial pigments. Organoleptic changes are those related to smell or taste. Examples include a sour taste due to the production of fatty acids; fishy smells caused by amines and the smell of rotten eggs due to sulphurous compounds. Changes in viscosity may also occur and result in a runnier product due to breakdown of thickening agents or perhaps even a more viscous solution due to the production of microbial polymers. Even if these events do not result in harm to the patient they will lead to loss of patient confidence and probably to noncompliance.

More profound microbial growth may lead to product deterioration and ultimately a nonfunctional medicine. Microbial metabolism of product ingredients can cause pH effects with acid or basic metabolites and the potential for secondary growth. Multiphase systems such as emulsions and creams may crack or become gritty. Bottles may even explode due to the build up of gases. There are a number of examples in the past where the presence of microbes has led to the degradation of formulation components such as excipients and actives.

17.6.2 Methods for the protection of products from microbial spoilage

From the above it is evident that the formulator/manufacturer must make every effort to ensure that their product is protected from microbial contamination. There are a range of methods that can be adopted.

17.6.2.1   pH

Figure 17.7 indicates that at the extremes of the pH range microbial growth is unlikely. There is however, still a wide pH range over which microorganisms (either fungi or bacteria) may grow freely. The use of extremes of pH as a means of protecting products has some value in the food industry such as with pickles and also with some household cleaning products, but for pharmaceuticals it is not such a useful mechanism.

17.6.2.2   Water activity (Aw)

Microorganisms cannot grow without water and so this presents a convenient way of controlling microbial contamination. Items such as tablets tend to be safe from microbial growth simply due to their dryness and often do not require further protection. Similarly, we do not need to worry unduly about nonaqueous products such as oils and ointments. However, it must be cautioned that it is not easy to predict what changes may occur to the product during its use. Storage in high-humidity environments may result in water ingress and negate these protective effects.

The important feature here is water activity, which is essentially the amount of uncomplexed water available to the microorganism for growth. Syrups, for instance, contain high amounts of water and yet most of it is unavailable to the microorganisms for growth. Water activity is the ratio of the vapour pressure of the product and the vapour pressure of pure water. Thus pure water has a water activity of 1. The water activity of other products is shown below:

A_w for most creams	0.8–0.98
A_w for Syrup BP	0.86
A_w for jam	0.7

Different microorganisms have limits of water activity below which they will not grow. The list below shows that Gram-negative bacteria require much more water to grow while fungi can more easily contaminate low water activity products.

Gram-negative bacteria	0.95
Gram-positive bacteria	0.90
Yeasts	0.88
Osmotolerant yeasts	0.73
Some filamentous fungi	0.61

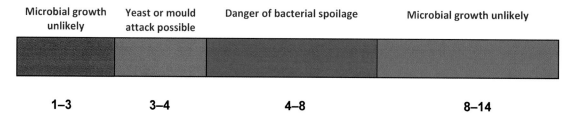

Microbial growth unlikely	Yeast or mould attack possible	Danger of bacterial spoilage	Microbial growth unlikely
1–3	3–4	4–8	8–14

Figure 17.7   The likelihood of microbial growth at different pHs.

17.6.3 The use of specific chemical preservatives

For the majority of multiple-use pharmaceutical products the only way to prevent microbial growth is by the use of specific chemical preservatives. Details of the individual biocides are given in Chapter 16 and will not be considered here. It is relevant however to consider the factors involved in the choice of these agents for particular products and how inappropriate formulation may lead to product failure. Table 16.1 in Chapter 16 identifies some of the formulation issues arising with the different groups of preservatives and indicates the types of formulations in which they might be used. Perhaps the most important point to make is that the inclusion of a preservative into a formulation should form part of the original design process and not just be added in at the end as an afterthought. Furthermore, a preservative should not be included to mask poor manufacturing processes or a deficient formulation.

An ideal preservative should have the following properties:

- broad spectrum of antimicrobial activity;
- rapid antimicrobial action;
- chemically stable and effective under all pH conditions;
- compatible with excipients and packaging materials;
- safe;
- cost effective.

Not surprisingly, there is no such thing as an ideal preservative and as a result we have to compromise in our choice of biocidal agent. When choosing an appropriate preservative a number of factors need to be taken into consideration with regard to its potential interactions with other formulation components. A few of these are considered below.

17.6.4 Factors influencing preservative activity

17.6.4.1 Specific chemical interactions

Cationic preservatives such as chlorhexidine and quaternary ammonium compounds may form insoluble products with inorganic anions and anionic surfactants. Benzoates and parabens form insoluble complexes with iron salts; chlorocresol is incompatible with phosphates and phenyl mercury salts are incompatible with chloride ions. In addition bronopol can complex with the aluminium present in collapsible tubes.

17.6.4.2 Partitioning

Poorly water-soluble preservatives such as parabens or the phenolics will tend to partition into any oil phase which may be present in the formulation. This is particularly problematic with multiphase products such as creams. As mentioned previously, it is the aqueous phase of the formulation which is at the greatest risk of contamination and hence it is this phase which must be protected. There is no point in having a formulation with the correct amount of preservative but where it is all concentrated in the oil phase which doesn't need protecting.

17.6.4.3 Adsorption to polymers

Many preservatives will bind to polymeric suspending agents such as tragacanth, alginate, starch mucilage and polyethylene glycols. In this state they are not available for antimicrobial action and so the product is essentially unpreserved.

17.6.4.4 Adsorption to suspended solids

Suspensions contain a high level of suspended solids and these can have a high surface area for adsorption. In the context of preservatives, parabens have been shown to adsorb to calcium carbonate and magnesium trisilicate; chlorhexidine adsorbs to kaolin and calamine, while benzoic acid can adsorb to sulphadimidine particles. This results in the same lack of available preservative as described above.

17.6.4.5 Physico-chemical effects

Phenolics, quaternary ammonium compounds and organomercurial preservatives are subject to photodegradation and so should be protected from light. Some agents such as the phenolics have a lower antimicrobial activity in low Aw solutions such as syrups and glycerol. Volatile preservative compounds may be lost from a product through evaporation from solution. Weak acids such as benzoic acid are only active at low pHs and if the pH drifts during storage the activity may change markedly. Table 17.3 shows the variation in ionization of sorbic acid at different pH values. The pKa of sorbic acid is 4.8, which means that at this pH the solution comprises equal amounts of ionized and unionized molecules. As the pH falls the proportion of unionized molecules increases and since this is the biologically active moiety (because of its ability to cross cell

Table 17.3 Variation in ionization of sorbic acid at different pH values.

pH	Ionized sorbic acid (%)
2	0.16
3	1.56
4	13.68
5	61.31
6	94.06
7	99.37

membranes) it follows that the activity is much higher at low pH values.

17.6.5 Reasons for product failure

In a number of the cases cited above a chemical analysis of the formulation as a whole may reveal a concentration of preservative, which, if used in a simple solution, would be sufficient to ensure effective preservation. However, since many of the molecules were taken from solution by adsorption or partitioning the amount of preservative remaining free in solution to provide an antimicrobial effect would be much reduced. Under these circumstances it is possible that the product might succumb to microbial contamination.

Other situations may arise during storage of the product leading to reduced preservative efficacy. In the example shown below in Figure 17.8 the preservative may adsorb onto the walls of the container and if the container is made of an appropriate polymer the molecule may actually absorb into the polymer and pass through to evaporate to the outside. This might arise if the original formulation was designed to be incorporated in a glass bottle for instance and, at a later date, this was switched to a plastic bottle for reasons of weight, safety or cost.

Figure 17.9 illustrates a situation where fluctuating storage temperatures might give rise to evaporation of water within the product and condensation on the lid of the container. This condensed water then falls back onto the surface of the product leading to a pool of liquid with reduced concentration of preservative and a higher water activity.

17.6.6 Is the loss of a small proportion of the preservative important?

In the examples cited above it may be that only a small percentage of the preservative in the formulation is lost or

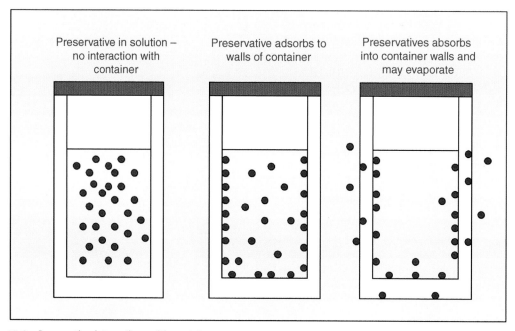

Figure 17.8 Preservative interactions with container.

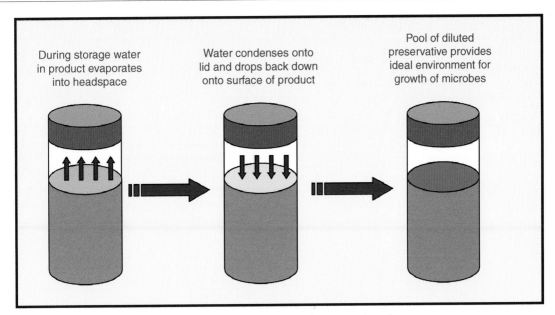

During storage water in product evaporates into headspace

Water condenses onto lid and drops back down onto surface of product

Pool of diluted preservative provides ideal environment for growth of microbes

Figure 17.9 The effect of fluctuating temperatures on the storage of a preserved product.

made unavailable. Hence, we might ask – does this really matter? The answer is that it might. Chapter 16 discusses the concept of concentration exponent where it is demonstrated that the loss of a small amount of preservative (if the preservative has a high concentration exponent – say in the case of phenol) will have a dramatic effect on biocidal activity.

In conclusion, therefore, a preservative must be designed into the formulation and not added as afterthought, nor should it be added to compensate for poor manufacturing processes or inadequate formulation. There is a need to understand fully how all the excipients will interact with the preservative and the effects this might have on the concentration of free agent. It is also vital to have a thorough knowledge of the concentration exponent of the biocide being used in order to determine the consequences of preservative loss.

Acknowledgement

Chapter title image: http://commons.wikimedia.org/wiki/File:Laboratoires_Arkopharma_Chaine_de_conditionnement_remplissage_piluliers.JPG

Chapter 18
The design of sterilization processes

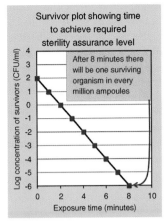

Survivor plot showing time to achieve required sterility assurance level

After 8 minutes there will be one surviving organism in every million ampoules

Calculating the numbers of surviving bacteria after heat treatment

KEY FACTS

- The word 'sterile' is an absolute concept. There are no degrees of sterility: an article is either sterile or it is not.
- It is neither possible to guarantee, nor to prove, that an article is sterile; it is possible only to quote a probability that it is sterile.
- This probability is referred to in the pharmacopoeias as the sterility assurance level and is usually quoted as 10^{-6}, meaning one surviving organism in one million items.
- The rate at which microorganisms are killed in a sterilization process is expressed as the D-value (decimal reduction time), which is the time required to reduce the population to 10% of its initial value.
- The inactivation factor quantifies the proportion or number of microorganisms killed by a sterilizing process.
- The Z-value is a parameter which indicates how killing rate is influenced by temperature; it is defined as the number of degrees Celsius temperature change required to achieve a tenfold change in D-value.
- F-values enable the killing effect of one autoclave cycle (temperature-time combination) to be compared with another. They also allow measurement of the extent of microbial killing during parts of the cycle – for example the heating and cooling phases; this may permit a reduction in the length of time that the autoclave needs to be maintained at the nominal sterilizing temperature.

The word 'sterile' is one that is easily misunderstood and, as a consequence, quite often misused. In a pharmaceutical context it describes an item (medicine, surgical device or dressing and so forth) that is *completely* free of living organisms. Thus, like pregnancy, it is an absolute concept; there can be no half-measures, an item is either sterile or it is not. Phrases like 'nearly sterile' or 'partly sterile' are as inappropriate as would be the phrase 'partly pregnant', and they simply display the user's lack of understanding. There is no level of contamination which is so low that it could be considered negligible and therefore acceptable; just one single surviving microbial cell would render an item nonsterile. Unfortunately, for reasons that will be explained later in this chapter, it is not possible to guarantee or prove that an item is sterile. This may lead to the belief that sterility is just a hypothetical

Essential Microbiology for Pharmacy and Pharmaceutical Science, First Edition. Geoffrey Hanlon and Norman Hodges.
© 2013 John Wiley & Sons, Ltd. Published 2013 by John Wiley & Sons, Ltd.

concept and the sterile state does not exist in reality; however, that is not so.

Because it is not possible to *guarantee* sterility, the best option is to quote a probability that an item is sterile. This has led to the concept of a 'sterility assurance level', which is usually quoted as 10^{-6}; in other words, there is a one in one million chance that the item in question is still contaminated. So, if a pharmaceutical manufacturer were to make a batch of an injection consisting of one million ampoules, one of them would contain a surviving organism – but there would be no way of identifying which one! In order to fully appreciate this probability basis of sterility it is necessary to understand the kinetics of microbial death.

18.1 Survivor plots and sterility assurance levels

When exposed to heat or radiation bacteria normally die according to first-order kinetics. This means that the same proportion of the cells is killed in successive time intervals, and it is illustrated by the hypothetical, simplified data in Table 18.1. The data could represent the death of bacterial spores in 1 ml ampoules of injection being sterilized by steam in an autoclave (also called a 'steam sterilizer'). The initial level of contamination in the liquid was 10^4 colony-forming units (CFU)/ml (which would be unrealistically high; presterilization bioburdens should normally be lower than this, but 10^4/ml is suitable for the purposes of this illustration). It can be seen from the middle column that in each successive one minute interval the spore concentration was reduced to 10% of the value at the start of that period.

Table 18.1 Simplified data illustrating bacterial death during steam sterilization.

Time of heat exposure	Viable spore concentration CFU/ml	Log$_{10}$ viable spore concentration
0	10 000	4
1	1000	3
2	100	2
3	10	1
4	1	0

If the concentration of surviving spores (middle column) is plotted against time the resulting graph is Figure 18.1A, in which the points would be concentrated towards the bottom of the y-axis. More commonly though, the logarithms of the concentration values are plotted, as in Figure 18.1B, and a graph of this type is known as a survivor plot or kill curve (despite the fact that when real experimental data are plotted there may be little or no curvature of the line).

Because the scales on both axes end in zero, it is tempting to regard Figure 18.1B as complete, with no possibility to extend the line for exposure periods longer than 4 minutes; however, this is not the case. Logically, the next line of data in Table 18.1 would be 5 minutes' exposure corresponding to 0.1 CFU/ml and a log value of -1. One-tenth of a viable colony forming unit sounds like a contradiction in terms because one-tenth of a bacterial spore could not be viable, but 0.1 CFU/ml corresponds to one whole spore in 10 ml of liquid. Given that this example describes sterilization of 1 ml ampoules of an

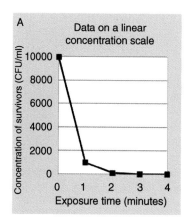
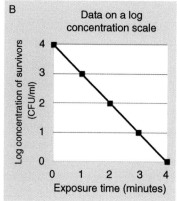

Figure 18.1 Survivor plots constructed from the data in Table 18.1.

injection, one surviving spore in 10 ml corresponds to nine of the ampoules being sterile and the tenth being nonsterile because it contains the survivor.

The next line in the data table would be 6 minutes' exposure corresponding to one survivor in 100 ml, then 7 minutes giving one survivor in 1000 ml, and so on. It is clear from this example that the data table could be extended indefinitely and zero survivors (sterility) would only arise after an infinite exposure time, so no matter how long the ampoules of injection are heated there is always a small but finite probability that there will be a surviving spore. This is the reason why it is not possible to *guarantee* sterility of an item selected at random from a sterilized batch of product. Instead, a sterility assurance level (SAL) of 10^{-6}, or better, is the target, and the exposure period in the autoclave is adjusted to achieve this because it can be calculated from knowledge of the presterilization bioburden and the degree of heat resistance of the spores (from their death rate – in other words, the slope of the survivor plot). So, in the example above, the required SAL (a log concentration of survivors of –6) would be achieved following exposure for 10 minutes (black squares in Figure 18.2). If, however, the presterilization bioburden were not 10 000 CFU/ml but 100 CFU/ml (which is a more realistic value) the plot would obviously start lower on the axis and the SAL would be achieved in 8 minutes rather than 10 (red triangles).

Good manufacturing hygiene that results in lower levels of microbial contamination prior to sterilization therefore affords several benefits:

- shorter autoclaving times that reduce energy costs;
- shorter heat exposure which is likely to reduce degradation of the active ingredient;
- fewer dead bacterial cells in the product after sterilization, which would reduce the risk of the injection failing the bacterial endotoxins test (see Chapter 3).

18.2 D-values

The slope of a survivor plot is obviously a measure of the organism's heat resistance, and it would be possible to express resistance by the slope. The equation representing the lines in Figure 18.2 is

$$\text{Log}_{10}N = \text{Log}_{10}N_o - \frac{kt}{2.303}$$

Where N = the concentration of survivors after t minutes' exposure, N_o is the initial concentration and k is the inactivation rate constant.

However, the numerical value of k, the rate constant, is rarely used because it is difficult to visualize how any quoted number translates into a killing rate. Much more commonly, the decimal reduction time (D-value) is used, and this is defined as the time required to kill 90% of the population. In Figure 18.3, for example, the D-value is exactly 0.75 minutes, so it is easy to understand that during every successive 45 second interval the population

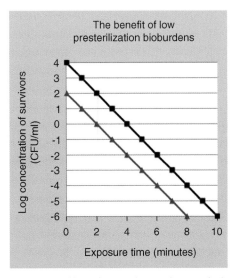

Figure 18.2 The effect of bioburden on the autoclaving time required for sterility assurance.

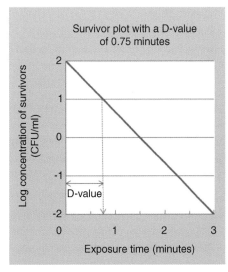

Figure 18.3 Survivor plot with a D-value of 0.75 minutes.

is reduced to 10% of its value at the start of that time, but the inactivation rate constant for the line is 3.07 min^{-1} – which is rather less informative. D-values are often quoted for bacterial spores exposed to steam-sterilizing temperatures, and in this situation a subscript is added to the D-value to indicate the temperature at which it was measured. D_{121} values are the most commonly encountered because 121 °C is the most frequently used autoclaving temperature; so, for example, spores of *Geobacillus stearothermophilus* (which is normally regarded as the most heat-resistant species) typically exhibit D_{121} values as high as 2.0 minutes or more.

Logically, D-values should only be quoted where the survivor plot is linear, and this situation exists for both heat and radiation sterilization but it is frequently not the case when bacteria are killed by toxic chemicals like disinfectants and antibiotics. When the viable population has fallen tenfold (in other words when the value on the y-axis has fallen by 1) it is often referred to as a decimal reduction, or a fall of one log cycle; so Figure 18.3 shows four decimal reductions.

18.3 Inactivation factor

The D-value quantifies the *rate* at which the bacteria are killed but, by itself, it does not indicate how many, or what fraction of the population, are killed; this is given by the inactivation factor (IF), which is calculated from the following equation:

$$IF = N_0/N = 10^{t/D}$$

where N_0 is the initial number, N is the final number, t is the exposure time and D is the D-value.

Thus the IF for *G. stearothermophilus* spores with a D_{121} of 1.2 minutes contaminating an injection which was being sterilized in an autoclave using the 'standard' cycle of 15 minutes at 121 °C would be:

$$IF = 10^{t/D} = 10^{15/1.2} = 10^{12.5}$$

The exponent, 12.5, means that the population would fall by 12.5 decimal reductions on a survivor plot like Figure 18.2. Knowledge of the bioburden and the inactivation factor permits calculation of the probability of survivors, and the inactivation factor can be used to show that different autoclave cycles may result in grossly different levels of microbial killing.

18.4 Z-values

Steam sterilization (heating in an autoclave) is by far the most commonly used terminal sterilization method, and the temperature-time combination of 15 minutes at 121 °C, which is suggested in the pharmacopoeias, is the heat treatment commonly used – but it does not have to be. A pharmaceutical manufacturer is not obliged to use this standard autoclave cycle; it is permissible to use any other temperature-time combination, provided that it achieves the required SAL. For example, 30 minutes at 115 °C was an autoclave cycle formerly recommended in the pharmacopoeias that has largely fallen into disuse now, and 3.5 minutes at 134 °C is currently used for the sterilization of some surgical dressings and instruments. It is logical to expect that the rate at which contaminating organisms are killed would increase as the steam temperature increased, so it would be useful, when designing a sterilization process, to have a parameter which made it possible to calculate by how much the rate was changed for a given change in temperature. Both temperature coefficients and activation energies (from Arrhenius plots derived from kill curves) will do this, but they are rarely used and will not be described here. The parameter that is almost invariably used in this situation is the Z-value, which is defined as the number of degrees Celsius temperature change required to achieve a tenfold change in D-value. This is illustrated by the following specimen data which would be typical for *G. stearothermophilus* spores.

Steam temperature (°C)	D-value (minutes)
99	120
110	12
121	1.2

In this example the Z-value is seen to be 11 °C, because the time required to kill 90% of the spores (D-value) is reduced tenfold as the steam temperature is increased from 99 to 110 °C, and again by a further tenfold from 110 to 121 °C. Z-values vary from one organism to another and are influenced by the experimental conditions, but when they are calculated from real heat resistance data they are often found to lie within the range 8–14 °C. Values close to 10 °C have been so commonly recorded that 10 °C is often assumed in sterilization calculations rather than measured from experimental data (see, for example F_0 values, below).

If both the Z-value and the D-value at one temperature are known, it would be possible to calculate the D-value at any other desired temperature from the equation below

$$Z = \frac{T_2 - T_1}{\log D_1 - \log D_2}$$

For example, if a population of spores with a Z-value of 10.5 °C had a D-value of 9.0 minutes (D_1 in the equation) at 115 °C (T_1 in the equation) and it was necessary to know its D-value at 121 °C (T_2 in the equation), the calculation would be:

$$\log D_1 - \log D_2 = \frac{T_2 - T_1}{Z}$$

$$\log 9 - \log D_2 = \frac{121 - 115}{10.5} = 0.571$$

$$\log 9 - 0.571 = \log D_2$$

$$0.383 = \log D_2$$

$$D_2 = 2.42 \text{ minutes at } 121°C$$

18.5 F and F_0 values

Two of the problems commonly encountered in the design of steam-sterilization protocols are:

- A variety of sterilizing temperature/time cycles have been recommended in the pharmacopoeias, yet they do not all afford the same degree of safety in terms of the extent of microbial killing. It would be useful to be able to compare one with another in terms of killing efficiency, and although inactivation factors can do this (see above) they require knowledge of the D-value for the most resistant bacterial contaminants of the product, and then a calculation, so an alternative easily understood parameter would be more useful.
- The standard temperature-time cycles are quoted as 'holding times' at the specified temperature, so, for example, the 121 °C, 15 minute cycle would satisfy the pharmacopoeial specification if the autoclave were capable of instantaneously heating from room temperature to 121 °C, remaining at that temperature for 15 minutes, then instantaneously cooling again – in other words, merely operating to give the central, shaded portion of the plot (labelled B) in Figure 18.4. Unfortunately, autoclaves do not work like that, and there are inevitable heating up and cooling down phases (labelled A and C respectively in Figure 18.4)

during which bacterial killing takes place. This killing, however, is surplus to the pharmacopoeial requirements, so it would be useful to have a parameter which calculated the killing effects during heating and cooling and would thus permit the 'holding phase' (labelled B) to be shortened proportionately. Again, this would afford cost savings in energy and minimize drug degradation.

F and F_0 values deal with both of the above problems by using 121 °C as a reference temperature and expressing the lethality of any autoclave cycle as the equivalent number of minutes' exposure at 121 °C. If for example, a cycle were stated to have an F value of 12 minutes, this would simply mean that it would achieve the same killing effect as 12 minutes at 121 °C.

To calculate an F value it is necessary to know the Z value for the spores, but since measuring it experimentally is time-consuming and the measured value is often found to be close to 10 °C anyway, this value of 10 is frequently assumed in F calculations. When this assumption is made the subscript zero is added to the F – it is designated F_0 – so, the pharmacopoeias define an F_0 value of a steam sterilization process as 'the lethality expressed in terms of the equivalent time in minutes at a temperature of 121 °C delivered by the process to the product in its final container with reference to microorganisms possessing a theoretical Z-value of 10'. F_0 values can be determined either by calculation or by use of special F_0 chart paper, but the latter is now rarely employed because computer software that can easily manipulate the data is so readily accessible.

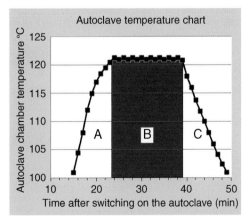

Figure 18.4   The change in temperature in an autoclave during a typical operating cycle.

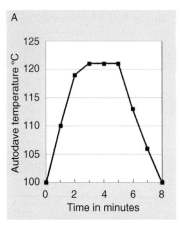

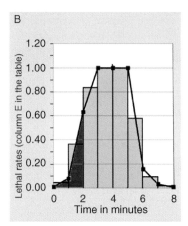

Figure 18.5 A. An autoclave temperature record (A). B. Converted to lethal rates (B).

The equation to calculate F_0 values is

$$F_0 = \Delta t. \Sigma 10^{(T-121)/Z}$$

This equation is easier to understand than it might look. It works by considering the entire autoclave cycle as if it were broken up into many short exposure periods, with Δt in the above equation indicating their duration. Normally, an industrial autoclave is fitted with multiple temperature probes (thermocouples) which record the temperature at different parts of the autoclave chamber or load, and from which the readings are transmitted through a flexible lead to a computer. Typically, the time interval between successive temperature readings is a fraction of a minute but if it were assumed to be exactly one minute then the term Δt would have a value of 1.

The remainder of the equation, the $10^{(T-121)/Z}$ part, calculates the lethal (killing) rate at a particular temperature, T, relative to the rate at 121 °C. If the killing rate during each time interval is calculated and multiplied by the time interval itself, the extent of bacterial killing during the interval can be calculated. If, for example, a temperature of 115 °C were maintained for one minute, the F_0 value would be:

$$F_0 = \Delta t \times 10^{(115-121/Z)}$$
$$F_0 = 1 \times 10^{(-6/10)}$$
$$F_0 = 1 \times 10^{(-0.6)}$$
$$F_0 = 0.251 \text{ minutes}$$

This result means that one minute's exposure at 115 °C has the same bacterial killing effect as 0.251 minutes at the higher temperature of 121 °C. If the

temperature were to be maintained steadily at 115 °C for 30 minutes then Δt in the equation would become 30 and the F_0 value for the whole period would be 30×0.251, which is 7.53 minutes. If the temperature changed throughout the autoclave cycle, which, of course, is the normal situation, it would be necessary to calculate the different killing effects during the various one minute intervals and add them together, and that summation is indicated by the symbol Σ.

A method for calculating an F_0 value using Excel is shown with simplified data below. Assume that the temperatures recorded at 1 minute intervals inside an autoclave were those in column B in Table 18.2 , which, when plotted, became Figure 18.5A. These values are converted to lethal rates according to the equation above by, successively, subtracting 121 from each temperature value (column C), then dividing by 10 (column D) and using these column D values as the power to which 10 is raised (column E). When time is plotted against the lethal rates in column E the resulting graph is Figure 18.5B, in which the total area under the curve is the F_0 value for the whole cycle. This total is calculated by adding together the areas under the curve during each one minute time interval, the first two of which are shaded red in the figure. Because most of these are triangle- or trapezium-shaped, the most convenient way of calculating the area is to take the mid-value between the plotted points and use this to create a series of rectangles which have the same area as their respective trapeziums (shaded grey and derived from column F in Table 18.2).

Temperatures lower than 100 °C have a negligible killing effect and are usually disregarded, so the relevant time period starts when the autoclave temperature reaches 100 °C.

Table 18.2 Excel spreadsheet tabulation for calculating F_0 values from autoclave temperature data.

A	B	C	D	E	F	G	H	I
Time	Temp°C	Column B	Column C	10 raised	Mid point			
minutes		minus 121	divided by 10	to power	between			
				in column D	values in E			
					(F_0 values)			
0	100	−21	−2.1	0.008				
1	110	−11	−1.1	0.079	0.044			
2	119	−2	−0.2	0.631	0.355			
3	121	0	0	1.000	0.815			
4	121	0	0	1.000	1.000			
5	121	0	0	1.000	1.000			
6	113	−8	−0.8	0.158	0.579			
7	106	−15	−1.5	0.032	0.095			
8	100	−21	−2.1	0.008	0.020			
					3.908	= F_0 values for the whole cycle		

The use of the F_0 concept in the design of steam-sterilization processes is described in the pharmacopoeias. The EP, for example, provides, in Section 5.1.5, several mathematical relationships between F_0 and other parameters described above, and emphasizes that 'When a steam sterilization cycle is chosen on the basis of the F_0 concept, great care must be taken to ensure that an adequate assurance of sterility is consistently achieved. In addition to validating the process, it may also be necessary to perform continuous, rigorous microbiological monitoring during routine production to demonstrate that the microbiological parameters are within the established tolerances so as to give an SAL of 10^{-6} or better.' What this means is that sterilization processes that are designed on the basis of F_0 calculations often have lower F_0 values and, consequently, involve less severe heat treatments than the 'standard' cycle of 121 °C for 15 minutes. It is, therefore, particularly important to ensure that the concentration and heat resistance of the contaminating organisms prior to sterilization are within predefined limits in order to avoid the risk of failing to achieve the sterility assurance level.

Chapter 19
Sterilization methods

A hospital- or laboratory-scale steam sterilizer (autoclave)

KEY FACTS

- Terminal sterilization is the preferred strategy for manufacturing sterile products because it is more reliable than aseptic manufacture.
- Fifteen minutes' exposure to steam at 121°C is the standard sterilization cycle described in the pharmacopoeias, but other temperature-time combinations may be used.
- High-temperature, short-time steam cycles give higher sterility assurance levels than low-temperature, long-time cycles; 133–135°C for 3–5 minutes is used for surgical instruments and dressings.
- Air must be removed from the autoclave chamber because mixtures of steam and air give substantially lower temperatures than pure steam at the same pressure.
- Glassware, heat-stable powders and oils may be sterilized by dry heat in hot air ovens or sterilizing tunnels at temperatures of the order of 160–180°C and 250–300°C respectively.
- Ionizing radiation and ethylene oxide gas are most commonly used to sterilize heat-sensitive materials, particularly medical devices containing plastics.
- Biological indicators – products containing spores of bacteria resistant to heat, radiation or ethylene oxide – are used to validate sterilization methods and, less frequently, to monitor them on a routine basis.
- Sterile filtration – in which solutions of heat-sensitive drugs are passed through bacteria-retentive filters – is frequently the only available method for drugs containing proteins, nucleic acids and carbohydrates.
- Tests for sterility are used as part of the sterility assurance process but they are unlikely to detect low levels of contamination in small manufacturing batches.
- Parametric release, in which a sterile product is released for sale or use without undergoing a sterility test, is permitted in certain circumstances.

Essential Microbiology for Pharmacy and Pharmaceutical Science, First Edition. Geoffrey Hanlon and Norman Hodges.
© 2013 John Wiley & Sons, Ltd. Published 2013 by John Wiley & Sons, Ltd.

19.1 Choice of method for manufacturing a sterile product

Two strategies are available for manufacturing sterile products: terminal sterilization, in which the product is made, packed in its final container, then sterilized; or aseptic manufacture where the product is made from individual sterile ingredients using aseptic techniques. Both approaches will be considered in more detail in this chapter, but it is worth noting that terminal sterilization, though less commonly used than aseptic manufacture, is still the preferred option because it is more reliable. The terminal sterilization methods available include the use of heat (either as steam or hot air), radiation and microbicidal gases, but none of them is universally applicable to all types of product, nor does any single technique fulfil all the following desirable properties of a sterilization method:

- reliable in terms of achieving the required sterility assurance level of 10^{-6};
- safe for the operators;
- safe in terms of inducing no damage to the product or its container, or inducing the formation of toxic residues;
- an easily understood process that can readily be controlled and monitored by physical instruments;
- short exposure time;
- low cost.

Heating methods are preferred by the World Health Organization and the pharmacopoeias, but many products, particularly medical devices containing plastics, cannot be heated, so radiation and ethylene oxide gas are used as alternatives. Bacterial spores are the most heat resistant of all organisms, so heat sterilization processes are designed with the aim of killing spores. If that is achieved, it is safe to assume that all other organisms – fungi, protozoa, viruses and non-spore-forming bacteria – will also be killed. The same principle applies to other sterilization methods too, such as those using radiation or microbicidal gases like ethylene oxide. Unfortunately, though, this logic does not work with prions (which are not living organisms, but are nevertheless infectious agents). They are more resistant than bacterial spores and have such an exceptionally high heat and radiation tolerance that sterilization processes designed to inactivate them would be likely to do severe damage to the product being sterilized. Indeed, with heat and radiation particularly, a compromise usually has to be made between killing the contaminating organisms

and damage to the active ingredient of the medicine, so enhanced safety (in terms of a greater sterility assurance level) can usually only be achieved at the expense of product damage.

Sterile filtration is a possible approach for heat-sensitive water-soluble drugs, but for sterile medicines that cannot be filtered, such as ophthalmic creams, aseptic manufacture may be the best option, although radiation is becoming more frequently used as a result of regulatory pressure to adopt terminal sterilization processes. The uses and operating conditions for the sterilization methods are shown in Table 19.1.

19.2 Steam sterilization

This is the most commonly used terminal sterilization method and, together with ionizing radiation, the most reliable. The autoclaves used for steam sterilization vary in size from benchtop models of the type commonly found in dental surgeries to large floor-standing industrial machines that are loaded with forklift trucks. Autoclaves used in quality assurance laboratories and hospitals typically have chambers of $0.2–0.5\,\text{m}^3$ – about the capacity of a large refrigerator. Benchtop and smaller, floor-standing machines (see, for example, Figure 19.1) usually generate their own steam from purified water added before the sterilization cycle starts, whereas the larger models are usually supplied with steam from an external boiler (Figure 19.2).

Autoclaves can have circular or rectangular chambers, be top-loading or front-loading and have doors at one end or, less frequently, at both ends of the chamber (to permit transfer of materials from a nonsterile to a sterile area for example). The doors of modern autoclaves are fitted with thermal locks which prevent the door being opened until the chamber contents are below a preset temperature, for example 80 °C; this affords greater operator safety but substantially increases the cycle time compared with older machines without this facility. Some autoclaves have an external jacket surrounding the chamber; this may be filled with steam to afford insulation to the chamber itself which, in turn, may result in faster heat-up times.

Steam is very much better as a sterilizing agent than water at the same temperature, because steam has a high latent heat content which is transferred to the objects being sterilized when the steam condenses on them. Consequently, steam quality is particularly important: it should be dry (containing no liquid water droplets) and saturated (containing the maximum amount of water vapour possible for the given temperature). Superheating

Table 19.1 Typical operating conditions and applications for common sterilization methods.

Method	Typical^a conditions	Common applications
Steam (heating in an autoclave)	Dry, saturated steam at 121 °C for 15 minutes	• Aqueous solutions in sealed containers (bottled fluids) • Surgical and dental instruments • Dressings (porous loads) • Decontamination of infected materials or laboratory waste
Dry heat (hot air oven)	160 °C for two hours, or, in a combined sterilization and glass depyrogenation cycle, 250 °C for 30 minutes	• Glassware • Oils, fats and waxes, and oily injections
Ionizing radiation	An absorbed dose of 25 kGy	• Heat sensitive (thermolabile) raw materials • Medical devices
Ethylene oxide	400–1000 mg/l of ethylene oxide, at 45–65 °C and 40–80% relative humidity for 0.5–10 h	• Medical devices
Filtration	Passage of liquid through a bacteria-retentive membrane with a pore size of 0.22 μm or less	• Solutions of heat-sensitive, water-soluble and oil-soluble materials

^{a}Referred to in the BP as reference conditions.

should be avoided; this occurs when the steam is not in equilibrium with the water from which it was generated, for example where the steam pressure is rapidly reduced without a corresponding fall in temperature, or where the pressure is kept constant but the steam temperature is artificially raised (because this would mean the steam was no longer saturated). Steam releases its latent heat on contact with the cooler items to be sterilized in the autoclave, but superheated steam will not do this until it has cooled to the 'correct' temperature so it behaves like hot air and is less efficient than saturated steam. High-temperature, short-time cycles (for example, 134 °C for 3–5 minutes) may afford the double advantage compared with lower temperature, longer time cycles (for example, 121 °C for 15 minutes) of causing less degradation of the active ingredient whilst providing a greater level of sterility assurance because of the higher inactivation factors achieved. Despite this, the latter is much more commonly used because it is the reference cycle quoted in the pharmacopoeias.

It is essential that air is removed from the autoclave chamber and completely replaced by steam during the operating cycle. Failure to remove all the air results in the temperature being lower than that for pure steam at the same pressure (lower than it should be), so

there would be a risk of sterilization failure. The correspondence between temperature and pressure in the autoclave chamber is a good indication of steam quality. The relevant SI unit of pressure is the kilopascal (kPa), although even modern autoclaves do not necessarily use this scale. Instead, the gauge might display pressure in the units of pounds per square inch (psi) or bar (a unit of pressure equal to 100 kPa) see Figure 19.3.

Pure steam at 15 psi (103 kPa) has a temperature of 121 °C, and any residual air in the chamber will result in a temperature lower than this. Data showing correspondence of temperature and pressure in accordance with published steam tables contributes to the assurance of sterility because it demonstrates that the process was operating correctly. Larger autoclaves may have vacuum pumps to remove the air, but many smaller ones allow the air to escape through a vent valve. Steam is lighter than air so it rises to the top of the autoclave chamber, and as it accumulates the air is gradually pushed down and out through the valve; autoclaves that remove air in this way are referred to as gravity-displacement or downward-displacement autoclaves.

It is essential that the item to be sterilized is fully hydrated, because steam kills microorganisms by

Figure 19.1   A small top-loading laboratory autoclave.

Figure 19.2 A front-loading laboratory autoclave having approximately four times the capacity of that in Figure 19.1. Note the thermocouple leads inside to record the temperature at different locations in the load.

- Validation of a new autoclave begins at the planning stage when the machine's specification is being considered and continues throughout its installation, commissioning and subsequent testing.
- Physical monitoring of the machine's performance in terms of pressure records and temperature data from multiple thermocouples would be required both for the

hydrolysing vital macromolecules; if there is insufficient water available the hydrolysis is less efficient. Consequently, vacuum-assisted autoclaves are required for the removal of air from dressings or from surgical and laboratory equipment in order for steam to penetrate to all parts of the load. On the same logic, autoclaves should not be used to sterilize the inside of sealed, empty, glass bottles because the organisms inside are simply being exposed to hot air, not to steam. A further benefit of a vacuum pump is that it can be used to dry the load at the end of the cycle – particularly useful in the case of dressings.

The validation of a new autoclave or a sterilization process for a new product is, itself, a subject upon which substantial book chapters have been written, so there is not scope here to provide all the details but the following points are fundamental:

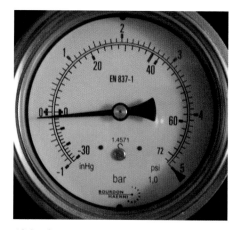

Figure 19.3 A pressure gauge on a modern vacuum-assisted autoclave. The red scale is in bar (see text) and the black scale above zero is calibrated in pounds per square inch. Note that the (black) vacuum scale (below zero) is in a third unit: inches of mercury.

autoclave operating empty and fully loaded with the materials it would be used to process.

- The load configuration is important because that may have a bearing on the steam penetration, so any change to the configuration would necessitate revalidation.
- Biological indicators (see Section 19.7) are also used for validation to confirm that the autoclave is capable of achieving sterility in the worst case scenario – one where the load is contaminated with high numbers of the most heat-resistant organisms; they would be placed free in the chamber and within the load itself.
- Revalidation is normally undertaken at regular intervals to confirm the continued safe operation of the autoclave, but any change to the product formulation, container or packaging or the operating conditions of the autoclave would necessitate earlier revalidation.

19.3 Dry heat sterilization

This method simply involves heating the item to be sterilized in a hot air oven, typically for 2 hours at 160 °C, but again, alternative combinations are available, for example 1 hour at 170 °C or 30 minutes at 180 °C. The temperatures and times required are longer than those for steam for two reasons: (i) because dry heat kills microorganisms by oxidative processes which are less efficient than the hydrolytic mechanisms of steam and (ii) because dry air does not possess latent heat. Items that can be successfully sterilized by dry heat (Table 19.1)

are, obviously, those that do not contain water, such as metal instruments, glassware, oils, fats and waxes or any heat-stable nonaqueous product like oily injections, ointments or powders. The method is also used for the depyrogenation of glassware required for the preparation or packaging of injections, but bacterial endotoxins (pyrogens) are very heat stable, so temperatures well in excess of 200 °C are necessary to achieve convenient exposure times; 250 °C for 30 minutes is common, though higher temperatures are used with shorter times.

Ovens used for laboratory-scale dry heat sterilization (Figure 19.4) are superficially similar to fan-assisted domestic kitchen ovens though usually with 2–3 times the capacity, but they differ by having temperature recorders, thermocouple inlets, thermal locking doors with noninterruptible operating cycles and the facility to filter the air that is drawn into the oven as it cools.

When dry heat is used on an industrial scale, sterilizing tunnels employing infrared irradiation or forced convection are more common. These operate continuously, and conveyor belts pass the items to be sterilized through a zone of hot air at 250–300 °C, with exposure periods of just a few minutes.

Items to be sterilized by dry heat need to be appropriately wrapped or sealed in containers that prevent recontamination after processing, and thermocouples are normally incorporated into the load in locations that would be expected to be slowest in achieving the desired operating temperature. Because both heat-up and cool-down times can be several hours, the full cycle

Figure 19.4 A laboratory-scale hot air sterilizer. *Source:* http://commons.wikimedia.org/wiki/File:Sterilisator_offen.jpg.

time from loading to emptying an oven can be a working day or more despite the fact that the 'holding' time may be just 2 hours or less.

19.4 Radiation sterilization

Gamma-rays or high-energy electrons are used to sterilize heat-sensitive materials and products like medical devices (for example, prostheses, intravenous giving sets, syringes and sutures), surgical instruments, anhydrous medicines (such as ointments) and powders. Gamma radiation is rarely used for water-containing products because the products of radiolysis of water usually cause too much damage to the drug and, even in the case of medical devices, damage to plastics in terms of discolouration and brittleness is a potential problem.

Ionization of atoms or molecules by gamma rays and high-energy electrons occurs without inducing radiation in the exposed material. It should be distinguished from nonionizing radiation, for example ultraviolet (UV) light, which possesses sufficient energy merely to produce molecular excitation, although this, in itself, may be sufficient to cause cell damage and death so UV light is used for surface disinfection. The unit for absorbed radiation dose is the gray (Gy) which has replaced the older unit of the rad (which equalled 0.01 Gy). In the United Kingdom, the standard radiation dose recommended for sterilization of pharmaceutical products is 25 kGy (2.5 Mrad in older textbooks) and it is worth noting that, as the complexity of living organisms increases, their susceptibility to ionizing radiation is increased, so the dose necessary to kill microorganisms by radiation sterilization, 25 kGy, is approximately 2500 times the lethal dose for a human (10 Gy or less). The mechanism by which radiation kills cells is that of ionization causing free radical production and damage to the DNA, although both human and microbial cells have damage-repair mechanisms.

Gamma rays, which are the more commonly used means of radiation sterilization, are produced from fuel rods normally containing ^{60}cobalt pellets. This isotope has a half-life of 5.25 years, so the rods have to be regularly replaced on a rotational basis to maintain the source strength. In contrast, high-energy electrons are generated electrically so no radioactive source is required. In both cases the items to be irradiated are passed by the radiation source on a conveyor belt or monorail in such a way that all sides of the object are irradiated. Table 19.2 compares the two methods.

Both gamma rays and electrons are more effective when oxygen is present and their activity increases with an increase in temperature although they do not, themselves, normally induce any significant temperature rise in the product being irradiated. Radiation sterilization facilities are expensive to construct and to operate, so companies manufacturing products to be sterilized in this way normally send them to one of a small number of specialist contractors. Health and safety considerations are paramount at radiation processing sites so thorough

Table 19.2 The relative merits of gamma and electron beam irradiation.

	Gamma	Electron beam
Source	Cannot be switched off, so it needs to be contained in a building having reinforced concrete walls up to 2 m thick and submerged in a storage pool of water at least 6 m deep when not in use	Can be switched off, so it poses no radiation risk when not in use
Speed	Relatively slow: may require several hours' exposure	A sterilizing dose may be given in minutes or even seconds
Penetration	Good: relatively large items (of the order 1 m^3) may be processed	Relatively poor so it is unsuitable for sterilizing dense materials, particularly medical devices containing metals
Product damage	Relatively long exposures can cause unacceptable product damage	Usually less damage than with gamma
Environment and public acceptability	• Spent fuel rods have to be disposed of • Public concern about radioactive materials	These problems do not arise

staff training and monitoring of radiation doses received by personnel are essential.

Ultraviolet light is a nonionizing form of electromagnetic radiation that has even poorer penetrating power than accelerated electrons. Light of a wavelength of 260 nm is the most effective because that is near to the absorption maxima for nucleic acids. It is commonly used for the disinfection of surfaces in aseptic work areas, air (as in microbiological safety cabinets and operating theatres for example) and for decontamination of water to be used both as an ingredient of medicines and for cleaning purposes. However, endotoxins, as well as other components or products of microbial cells like antigens and enzymes, are likely to remain even after the cell has been killed, so UV treatment could not be used to produce endotoxin-free water for injection.

19.5 Gaseous sterilization

Several microbicidal gases have been used for sterilization including ethylene oxide, formaldehyde, hydrogen peroxide and peracetic acid but, of these, ethylene oxide (sometimes referred to as EtOx) is by far the most common and will be the only one considered here. It is not a favoured method because it is less reliable than heat and radiation, so it needs rigorous in-process monitoring to confirm that sterilizing conditions have been achieved; it is also slow and there are several safety issues concerning its use, so it is only employed when there is no alternative. Its use is rather more common in the United States than in the United Kingdom and Europe.

Ethylene oxide is suitable for sterilizing materials that are both heat and radiation sensitive, so it is used primarily for disposable medical devices. It is also infrequently used in hospitals for surgical instruments and the sterilization of isolators and chambers, although hydrogen peroxide is now preferred. Ethylene oxide diffuses easily into paper, rubber and many plastics, but it cannot easily penetrate into crystalline materials and its activity is significantly reduced by the presence of organic material (blood, pus or faeces) so it cannot be used to sterilize crystalline raw materials or unwashed surgical instruments. Hospital ethylene oxide sterilizers are similar to conventional autoclaves, being steel chambers of varying capacities from about 65 litres upwards, whilst industrial-scale sterilizers are very much larger (Figure 19.5)

In contrast to heat and radiation methods, there is no single widely adopted set of sterilizing conditions for ethylene oxide. Its activity increases with temperature, humidity and gas concentration, so these parameters may be varied generally within the ranges shown in Table 19.1

Figure 19.5 Loading an ethylene oxide sterilizer. *Source:* Image courtesy of Steris Corporation.

with exposure periods varying from 1 to 10 hours although 3 to 4 hours is typical. Because materials absorb the gas so readily it is common for the concentration to fall during the cycle, so additional gas, and, possibly, steam, may need to be introduced into the sterilizing chamber during the cycle. It is normal for the materials being sterilized to be prehumidified before the gas exposure begins.

Ethylene oxide is a colourless gas that is explosive when mixed with air in proportions greater than 3.6% by volume, so it is normally used as a mixture with carbon dioxide (8.5–80% of ethylene oxide), nitrogen or dichlorodifluoromethane (12% ethylene oxide) to minimize the risk. Alternatively it is introduced into an evacuated sterilization chamber as the pure gas at subatmospheric pressure. The sterilization cycle starts with an initial vacuum to remove the air, followed by preliminary heating and humidification with pulsed subatmospheric pressure steam, which would be removed by a second vacuum prior to admission of the ethylene oxide gas into the chamber. After exposure is complete the gas is pumped out and replaced by sterile, filtered air and, usually, the materials are then aired for several hours or even days before use, in order for the absorbed ethylene oxide to dissipate.

Ethylene oxide is thought to kill microorganisms by alkylating essential proteins and nucleic acids in the cell; this mechanism of action means that the gas is both mutagenic and carcinogenic. It also causes acute eye, skin and bronchial irritation at concentrations above 200 parts per million (ppm) but, crucially, many people are unable to detect it by smell until the concentration is about three times that value, or more. Health and safety aspects of ethylene oxide sterilization are therefore a major consideration. In addition to the long cycle times and the need for extended airing, the other drawbacks are its inability to inactivate pyrogens and the requirement for the routine use of biological indicators in every load (not just for initial validation) and all loads being subjected to sterility tests (see below). Against that, the method does afford the advantages of being well recognized by regulatory authorities and causing only a low incidence of product damage.

19.6 Filtration sterilization and aseptic manufacture

Many drugs and pharmaceutical materials are damaged both by heat and radiation and are not suitable for gaseous sterilization either. Advances in genetic engineering and other aspects of biotechnology have resulted in the introduction of a large number of drugs containing proteins (for example, monoclonal antibodies, interferons) and other biological polymers (such as polysaccharide- or DNA-vaccines), and for many of them the only sterilization option is to use filtration to physically remove the contaminating microorganisms and then manufacture the product aseptically using sterile ingredients. However, filters are available for the removal not only of bacteria and larger organisms like yeasts and mould spores, but viruses too, and they can even remove pyrogens (endotoxins), so filtration is a common means of sterilizing injections and eye drops as well as air and other gases. It should be emphasized, though, that filters designed specifically for endotoxin or virus removal are not routinely used for sterilization purposes. An incidental advantage of filtration sterilization is that pharmacopoeias set limits for the number of particles (of fibres, dust and so forth) that are permissible in injections, and so sterilizing filters, which, of course, remove nonmicrobial particles too, do two jobs in one.

In the pharmaceutical industry, the most common materials used for modern sterilization-grade filters are cellulose derivatives and polymers like PTFE, polycarbonate and polyethersulfone. These can all be made as flat sheets, which are often circular so that they fit into purpose-made cylindrical holders, or are pleated to increase their surface area. Commonly they are made into cylindrical cartridges. The advantage of using synthetic polymers is that the pore diameter and density (pores per square millimetre) can be quite precisely controlled; this is important not just from the perspective of removing particles but because it markedly affects the flow rate of the liquid through the filter. Filter units can vary considerably in size from industrial-scale cartridges a metre or more in length to miniaturized disposable

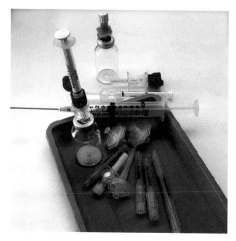

Figure 19.6 A selection of disposable units that fit onto syringes for sterilizing small volumes of liquid *Source:* Image courtesy of Helapet Limited.

syringe filters (Figure 19.6) that can be used for sterilizing small volumes of drug solutions, such as eye drops.

A distinction is made between so-called depth filters and membrane filters. The former are more likely to be made from ceramics, glass or metal and have a significant thickness relative to their pore size, whereas a membrane filter, normally of a synthetic polymer, is comparatively thin, although there are no well defined size rules for distinguishing them. These two categories fulfil, to different degrees, the characteristics of an ideal filter: both afford good particle removal (sterilizing efficiency) and mechanical strength and they are easily sterilized *in situ* by steam, but the membrane filter affords the additional advantages (not possessed by depth filters) of low fluid retention, low solute absorption, no grow-through of microorganisms and no shedding of fibres into the filtrate. Depth filters on the other hand have a high dirt-handling capacity and operate for a long period before clogging; for this reason they are often used as prefilters in front of a membrane filter to extend its life. They cannot easily be compared in terms of speed of filtration because this depends on the characteristics of the filter material itself.

Filters are available having a wide variety of pore sizes, and for sterilizing purposes 0.2 μm or 0.22 μm diameter pores are recommended (Figure 19.7) but 0.1 μm membranes are becoming more popular.

Sieving is one of the principal filtration mechanisms by which microorganisms and other particles are removed from solution and it is considered to be the most reliable. Adsorption of particles onto the surface of the filter or the sides of the pore channels is also a significant mechanism, however, as is the simple

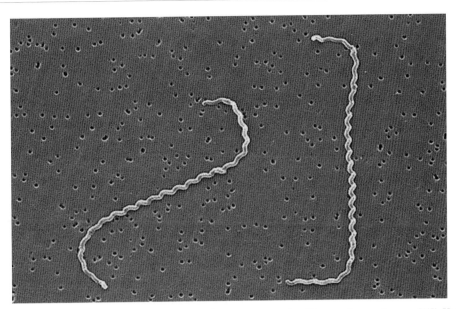

Figure 19.7   Leptospira bacteria attached to the surface of a 0.2 μm pore size filter membrane. *Source:* PHIL ID #1220; Photo Credit: Janice Haney Carr, Centers for Disease Control and Prevention.

entrapment of particles within these tortuous channels. The largest common viruses are about 0.3 μm diameter, and go they down in size to less than one-tenth of that value, so it is clear that sieving by a 0.2 μm filter is unlikely to be an effective means of removal of the smallest viruses. Instead, virus-retentive filters are available, which have nominal pore sizes of 0.01–0.02 μm, although pore size is a less reliable guide to the efficiency of viral removal than the log reduction value (LRV), which, as the name suggests, is the logarithm of the ratio of the number of organisms challenging the filter to the number passing through it; LRV values of 3 to 6 are common but there is often a tradeoff between viral reduction and adsorption of the active ingredients of the solution, particularly proteins.

An advantage afforded by filtration is that the method physically removes both living and dead cells from solution and in this respect it differs from heat and radiation methods where the dead cells remain in the product and possibly contribute to the pyrogen load. Thus a filter-sterilized solution would be expected to have a greater probability of passing a pyrogen or endotoxin test anyway, but this can be further enhanced by the use of positively charged polyvinylidene fluoride or nylon filters, which attract the negatively charged endotoxins.

Filtration is the most convenient method of sterilizing air and other gases and it is used to supply air to pharmaceutical manufacturing suites ('clean rooms') and isolators, operating theatres and microbiological safety cabinets. Depth filters, typically made of glass microfibres separated by aluminium sheets, are normally used for gas filtration and they attract particles of different sizes by the interception, impaction and diffusion mechanisms shown in Figure 19.8. High efficiency particulate air (HEPA) filters typically remove 99.97% of airborne particles of 0.3 μm in diameter, although some with higher specifications exist.

There is no simple method of monitoring the operation of a filtration sterilizing process as there is with heat and radiation, although a sudden fall in pressure on the upstream side of the filter might indicate a fracture of the filter itself. Representative filters from a manufactured batch are subjected to a microbial challenge test using *Brevundimonas (Pseudomonas) diminuta* as described in Table 19.3, but this is a destructive test that could not be applied to a filter intended for the manufacture of a medicine; instead, bubble point and diffusive flow tests are performed on filters prior to use. The former, which is less frequently used, measures the pressure required to cause bubbles of gas to emerge through a water-covered filter, whereas the latter determines the pattern of gas diffusion through a filter at progressively increasing pressures.

It should be emphasized that filtration is not a terminal sterilization process, defined as one in which the product is sealed then sterilized in its final container. Solutions that are filter sterilized still have to be dispensed into their containers and, despite the operation being undertaken in a class A atmosphere (see Chapter 17), there is the opportunity for contamination to arise during this process.

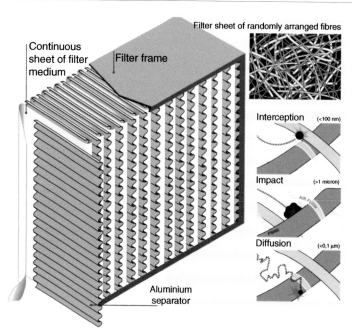

Figure 19.8 Cross-section of a HEPA filter showing three different mechanisms of particle removal. *Source:* http://commons .wikimedia.org/wiki/File:HEPA_Filter_diagram_en.svg.

Table 19.3 Biological indicators of sterilization.

Sterilization method	Indicator organism named in the European Pharmacopoeia	Indicator specification	Pharmacopoeial performance characteristics
Steam	*Geobacillus stearothermophilus*[a]	In excess of 5×10^5 spores per carrier with a D_{121} of not less than 1.5 minutes	$121 \pm 1\,^{\circ}$C for 6 min leaves revivable spores but $121 \pm 1\,^{\circ}$C for 15 min leaves no survivors
Dry heat	*Bacillus atrophaeus*[a]	In excess of 1×10^6 spores per carrier with a D_{160} of not less than 2.5 min	Not specified
Ionizing radiation	*Bacillus pumilus*	In excess of 1×10^7 spores per carrier with a D value of not less than 1.9 kGy	No survivors after exposure to 25 kGy
Ethylene oxide	*Bacillus atrophaeus*[a]	In excess of 1×10^6 spores per carrier with a D-value not less than 2.5 min when exposed to a test cycle involving 600 mg/l of ethylene oxide, at 54 $^{\circ}$C and at 60% relative humidity.	No survivors after 25 minutes' exposure to 600 mg/l at 54 $^{\circ}$C and at 60% relative humidity, but 600 mg/l at 30 $^{\circ}$C and 60% relative humidity leaves revivable spores
Filtration	*Brevundimonas diminuta*[a]	The filter must be capable of retaining a challenge of at least 10^7 CFU per cm^2 of active filter surface	

[a]The names of these organisms have been changed, so the former names will still be found in older textbooks: *Geobacillus stearothermophilus* = *B. stearothermophilus*; *Bacillus atrophaeus* = *B. subtilis* var *niger* and *Brevundimonas diminuta* = *Pseudomonas diminuta*.

To confirm the suitability of the environment for the purpose, sterile products' manufacturers are required to undertake process simulations (commonly known as 'media fills') in which the same factory filling line that would be used to fill the product in question is first tested by dispensing sterile culture medium into, typically, 5000–10 000 sterile containers. These are sealed and incubated, and should, ideally, yield none showing growth; one positive container would be deemed by the regulatory authorities to be sufficient to justify an investigation and two would require the whole process to be repeated. Because filtration is inherently less reliable than terminal sterilization processes, products sterilized in this way must be subjected to a test for sterility.

19.7 Biological indicators

Table 19.3 identifies the characteristics of the common biological indicators. For steam sterilization *Geobacillus stearothermophilus* spores are invariably used because of their extreme resistance, ease of cultivation and non-pathogenicity (they are thermophiles so they will not grow at body temperature). The spores are dried onto paper strips or other carriers made of glass, plastic or metal, or they are used as aqueous suspensions in sealed ampoules. After exposure during the autoclave cycle under test, the indicators are incubated in culture medium according to the manufacturer's instructions and absence of bacterial growth is taken to mean the spores have been killed and the sterilization cycle was satisfactory. Incubation typically lasts for one week so confirmation of the safety of the cycle cannot be given before then, and this time delay is an inconvenience. Consequently, there are a number of chemical indicators on the market which are intended to closely mimic the performance of the biological indicator; these usually operate on the basis of a colour change occurring after exposure to sterilizing conditions. They are useful for routine day-to-day monitoring in terms of supplementing temperature and pressure data but their use is not permitted by the regulatory authorities as substitutes for biological indicators during autoclave validation.

19.8 Tests for sterility

The principle of a test for sterility is quite straightforward: the item to be tested is placed into liquid culture medium and if, after incubation, there are no signs of growth (turbidity) the item is deemed to have passed the test. Actually conducting a test is not quite so simple, however,

because there are several controls required to ensure validity, and both operator technique and testing facilities need to be of the highest standard to avoid the introduction of microbial contamination during the test itself, which would give rise to a false positive result. Because of this, many companies, even quite large ones, do not conduct their own sterility testing but engage specialist contract laboratories to undertake it on their behalf.

From a financial perspective, sterility tests and endotoxin tests could be considered to be more important than many of the other quality-control tests to which a medicine is subjected because they are undertaken at the very end of the manufacturing process when all the added value has been built into the product. If a product were to fail a sterility test it is likely that the batch would be discarded, so it is particularly important that a sterility test is conducted with all possible care in order to maximize the chances of it giving the correct answer. Failing and scrapping a batch that is really sterile costs the money that has been invested in its manufacture, whilst passing a batch that is really contaminated risks initiating infections in patients who receive the medicine. Contaminated batches are likely to lead to product recalls, damage to the company reputation, major scrutiny from regulators and possibly even litigation.

Tests for sterility have been internationally harmonized, which means that the description of the test in the British Pharmacopoeia is the same as those in the European, United States and Japanese pharmacopoeias. It is not the intention here to describe in detail how the test is conducted because that information is given very concisely in the pharmacopoeias themselves. Rather, the aim is to explain the factors that may influence the reliability of the test and the interpretation of results.

To minimize the risk of contamination being introduced during the test, the pharmacopoeias recommend that sterility tests should be conducted in a class A laminar flow cabinet, which is, itself, located in a class B clean room. Increasingly, isolators (see Figure 17.5) are used as alternatives to laminar flow cabinets. There is also a caution about using operating conditions that might affect the viability of any microorganisms present in the product; this means, for example, that it should not be directly exposed to the ultraviolet lights or disinfectants which are used to decontaminate the cabinet or isolator before use. Regulators would expect there to be documented evidence of operator training and regular reassessment of competence.

Two culture media are recommended in the pharmacopoeias: fluid thioglycollate medium is primarily suitable for the growth of anaerobic bacteria, whereas soyabean casein digest medium (more commonly known as tryptone soya broth) is suitable for aerobic bacteria and fungi. Other media may be used provided that they are shown to be equally suitable, but in fact very few testing laboratories exercise this option. The two recommended media will support the growth of common organisms that might constitute the presterilization bioburden of many sterile products but it is important to recognize that there is no such thing as a universal culture medium, so there are organisms that could plausibly be present in an inadequately sterilized product that would not grow in, or be detected by, these media. Quite apart from the problem of these nondetectable bacteria, another fundamental flaw in sterility-testing protocols is that they do not check for the presence of viruses or other obligate intracellular pathogens like chlamydia and rickettsia which, because of their small size, are the organisms most likely to penetrate bacteria-retentive filter membranes used in aseptic manufacture and so, arguably, the organisms most likely to contaminate an inadequately sterilized product.

There are several controls that are incorporated in sterility tests:

- It is self-evident that the media used to detect the presence of any contaminants should, themselves, be sterile, so it is necessary to incubate unopened containers of media, all of which should show no signs of growth after incubation.
- There are fertility controls which are intended to confirm that the media do, in fact, support the growth of common contaminants from low inoculum levels.
- Controls to confirm that any antimicrobial activity present in the product (preservatives and antibiotics for example) is effectively neutralized. This is necessary to avoid the possibility of such antimicrobial chemicals reducing the growth rate of any contaminants so that their presence is not revealed as turbidity in the liquid within a standard 14-day incubation period. The best way of removing antimicrobial activity is to use the membrane filter method of sterility testing. This is officially preferred to the alternative, which is direct inoculation of the sample into the culture media. In membrane filtration, liquid samples are passed through $0.45\,\mu m$ pore-sized membranes in order to filter out contaminating organisms and the membranes are then washed with culture media (possibly containing specific preservative neutralizers) before being transferred into liquid media, incubated and scanned for signs of growth. It is important to recognize that the use of membrane filters is not here associated with the counting of bacterial colonies, as it is in bioburden determinations. The number of surviving organisms is

irrelevant because just one would be sufficient to cause the product to fail the test.

- Finally, the pharmacopoeias require the use of 'negative controls', which are intended to assess the adequacy of the facilities and the operator's technique.

The reliability of a sterility-test result increases with the number of samples taken for testing, but it is a destructive test so the more samples that are taken, the fewer there will be left to use or sell. To avoid the laboratory staff facing the dilemma of how many samples they should take, the pharmacopoeias specify precisely how many items should be tested for batches of different size. Even when the 'correct' numbers are used, however, there is still a substantial risk that the sterility test, despite being performed without fault, will fail to detect low levels of survivors in an inadequately sterilized product. It is possible to calculate the probability that the test will detect the organism(s) for any combination of sample size and true level of contaminant. For example, if 20 ampoules were taken for testing from a typical, large industrial-scale batch of injection and the true level of contamination was one ampoule in 100 containing a survivor, the probability of detection would be just 0.18: in other words, there would be an 82% chance of getting it wrong! Tables showing these probabilities for different batch sizes and contamination levels are published in other textbooks, but it is sufficient here to emphasize that sterility tests simply cannot reliably detect low levels of contamination no matter how well they are conducted.

A philosophical problem with sterility testing is that it is seeking to confirm a negative – trying to show that something is not there. So if the objects being sought, in this case contaminating organisms, are not found, the test is always vulnerable to the criticism that the procedure was not sufficiently rigorous to find them, and if the incubation conditions, or the media, or the sample size had been varied, then the presence of surviving organisms would have been demonstrated. Thus the test is merely one for the absence of gross contamination with easily grown organisms, and if a batch of product passes a test for sterility it does not provide absolute assurance that the batch is sterile. So, again, sterility is not claimed or guaranteed – the result is merely stated in terms of the batch 'having passed the test for sterility'.

19.9 Parametric release

Quite apart from the shortcomings of sterility tests in terms of their low probability of detecting small numbers of contaminating organisms, such tests also suffer from the disadvantages of a 14 day incubation period, during which time the product is in quarantine and cannot be released for use or sale, and the fact that even in the best testing facilities there is a small, but significant, incidence of false positive test results, typically 0.05–0.1%. However, when parenteral products, typically with low bioburdens of the order of 10 CFU per container, are sterilized by heat and radiation methods designed to achieve a sterility assurance level of 10^{-6} or better, it is clear that the technology of the sterilization process itself inspires much greater confidence of sterility than the sterility test, which was originally designed to assess its effectiveness.

For these reasons, it has been possible since the 1980s for manufacturers to apply to the regulatory authorities for permission to release products without sterility tests using a system termed 'parametric release'. The United States' FDA has defined this as a 'sterility-release procedure based upon effective control, monitoring and documentation of a validated sterilization cycle *in lieu* of release based upon end product sterility testing'; the same process is available through the European regulators. In both cases it applies to heat- and radiation-sterilized medicines and medical devices and cannot be used for filter-sterilized, aseptically manufactured products. In the United States, parametric release for ethylene oxide sterilized products is also possible.

Because the principle is that the sterilization process itself affords the assurance of sterility there are strict regulations to be observed. These include:

- The use of a process validated to achieve a 6 log cycle reduction in bioburden.
- Presterilization bioburden determinations on each batch of product that must meet predefined criteria and confirmation that any spore-formers present have a lower resistance level than the organisms used to validate the process.
- The use of physical, chemical or biological indicators in each load to confirm the effectiveness of the cycle.
- The use of a container and closure system that has been shown to prevent entry of microorganisms throughout the product's shelf life.

Acknowledgement

Chapter title image: PHIL ID #5753; Photo Credit: Troy Hall, Centers for Disease Control and Prevention. Microbiological waste being decontaminated in a large laboratory autoclave.

Chapter 20

The use of microorganisms in the manufacture of medicines

Alexander Fleming – discoverer of penicillin

KEY FACTS

- Microorganisms are used to produce a wide range of complex molecules, which cannot be made synthetically.
- Examples include enzymes, proteins, peptides, steroids and most antibiotics.
- Primary metabolites are produced by microorganisms during their period of active growth.
- Secondary metabolites are produced after the cell has stopped growing.
- The amount of material produced naturally is very small and microorganisms must be mutated to make them overproduce the product of interest.

The majority of chemicals used as medicines are manufactured synthetically with resultant high yield and purity. However, there are a number of instances where it might not be possible to produce the chemical in the laboratory and this is most commonly seen where the molecule is highly complex such as with steroids or proteins and peptides. Under these circumstances it might be more advantageous to use the manufacturing facilities within a microbial cell to carry out the complex synthetic processes. The molecule may then be extracted and, if necessary, modified further in the laboratory.

In fact there are a number of advantages to using microbes as synthetic factories:

- They have a high metabolic rate.
  - Microorganisms are small.
  - They have a large surface-to-volume ratio.
  - Rapid transport of nutrients into cell.
  - Cells grow very fast.
- They possess a wide range of enzymatic capability.
  - Fungi in particular are saprophytes and so live on dead and decaying matter.
  - They can produce many different enzymes.

Essential Microbiology for Pharmacy and Pharmaceutical Science, First Edition. Geoffrey Hanlon and Norman Hodges.
© 2013 John Wiley & Sons, Ltd. Published 2013 by John Wiley & Sons, Ltd.

- They produce an extensive range of metabolic end products.
- Microbial reactions are carried out under mild environmental conditions.
  - Many industrial chemical reactions require high temperatures, high pressures or the use of organic solvents – this makes the reactions expensive to carry out.
  - Microbial systems require low temperatures and pressures and usually employ aqueous solvents.
  - This results in lower energy costs.
- Microorganisms can grow on plentiful supplies of cheap nutrients such as the end waste products of other manufacturing processes.
  - Waste paper pulp.
  - Petroleum products.
  - Molasses.
  - Cornsteep liquor.

The summary box gives examples of a range of different products obtained from microorganisms and these will be discussed in more detail below.

Summary box showing typical products of microbial synthesis

- The cells themselves.
  - Baker's yeast/brewer's yeast used industrially and in domestic kitchens.
  - Quorn (mycoprotein from *Fusarium graminearum*) and edible mushrooms.
  - Marmite (yeast extract).
  - Probiotics.
  - Single cell protein for animal feed.
- Large molecules such as enzymes, polysaccharides and proteins (both natural and bioengineered). Tables 20.1 and 20.2 illustrate the range of pharmaceutical products that are able to be produced.
- Microbial biotransformations.
  - The main example here is the biotransformation of steroids.
- Primary metabolic products arise during active growth of the microbial culture and high cell concentrations and high growth rates give rise to high yields. The microorganisms used industrially are usually mutated to greatly overproduce the products of interest which include:
  - Alcohol.
  - Vitamins.
  - Amino acids.
  - Nucleotides.
  - Organic acids (including vinegar).

- Secondary metabolic products are produced after the cells have finished actively dividing and are in stationary phase. Conditions giving high growth rates give poor secondary metabolite yields, hence there is a need for slow growth rates but high cell concentrations. Examples of secondary metabolites include:
  - Antibiotics.
  - Toxins.
  - Alkaloids.

20.1 The cells themselves

Microbial cells are mainly used in the food industry and as feedstuffs for farm animals. Examples are given above and with the exception of probiotics they have little relevance in the pharmaceutical field and so we will not talk about them further.

20.2 Enzymes, proteins and polysaccharides

This group represents a wide range of products used in the pharmaceutical and cosmetic industries. It also includes a large number used in the food and chemical industries which do not have direct relevance here. Many of them are multibillion dollar markets including biological washing powders which contain microbially derived protease enzymes. Table 20.1 gives some examples of the different types of products which are of pharmaceutical interest. Recombinant proteins will also be briefly considered here.

20.3 Recombinant proteins

A number of pharmaceutically useful compounds of biological origin are not produced naturally by microorganisms but by human cells, and examples are given in Table 20.2. It is very difficult to culture human cells for the industrial production of these compounds and so other strategies have to be adopted. One such strategy is to identify the gene which produces the compound of interest, then to splice that into the genome of an easily grown microorganism such as *E. coli* and grow that organism in culture. Provided we have the mechanism to switch on the gene the bacterium will produce large quantities of the final product.

It is not the purpose of this book to give further details of this complex subject but the summary box on gene cloning outlines the basic process and the reader is referred to the website for further information.

Table 20.1 Examples of enzymes, proteins and polysaccharides produced by microorganisms for commercial purposes.

Material	Producing microorganism	Applications
Alginate	*Pseudomonas* sp; *Azotobacter vinelandii*	Used in a wide range of pharmaceutical formulations
Asparaginase	*Erwinia chrysanthemi*	Treatment of acute lymphoblastic leukaemia
Dextran	*Leuconostoc mesenteroides*	Blood plasma substitute
Hyaluronic acid	*Streptococcus zooepidemicus; Bacillus subtilis*	Wide range of pharmaceutical and cosmetic applications
Levan	*Zymomonas* sp.	Cosmetic applications
Streptokinase	*Streptococcus* sp.	Treatment of thrombosis in myocardial infarction and pulmonary embolism

Table 20.2 Recombinant proteins that are used clinically.

Recombinant protein	Therapeutic use
α-1 antitrypsin	Treatment of emphysema
Erythropoietin	Treatment of anaemia
Factor VIII	Prevention of bleeding in haemophiliacs
Granulocyte colony stimulating factor	Stimulates white blood cells, aids recovery of cancer patients from neutropenia and chemotherapy
Human β-glucocerebroside	Treatment of Gaucher's disease
Human growth hormone	Growth promotion
Human tissue plasminogen activator	Dissolves blood clots (acute myocardial infarction)
Insulin	Treatment of diabetes
Insulin-like growth factor	Growth promotion
Pegylated interferon 2a	Antiviral, antitumour
Tumour necrosis factor	Antitumour

Summary information on gene cloning

Human genes responsible for producing complex biological molecules such as hormones, enzymes and cytokines can be inserted into bacterial cells. These cells are easily grown to high cell densities in large volumes and the desired therapeutic materials produced on a large scale. There are three elements to the process:

- The gene of interest – derived from human DNA.
- A small piece of bacterial DNA, which can act as a carrier for the gene of interest (a plasmid vector).
- A host bacterium (typically *Escherichia coli*) into which the plasmid vector will be inserted to produce the biological product.

The human DNA
The human DNA is extracted and purified and then cut into small pieces using a restriction enzyme such as EcoR1. These enzymes recognize specific base sequences on the DNA and result in an uneven cut at the ends of the pieces of double-stranded DNA (called sticky ends).

(continued)

x

Plasmid vector

These are small circular pieces of double stranded DNA which have the capacity to replicate autonomously within the bacterial cell. They contain a small number of genes including an antibiotic resistance gene (often ampicillin) and a gene to allow screening for cells containing foreign DNA (this may be a gene for β-galactosidase). The plasmid is treated with the same restriction enzymes to generate sticky ends at the sites of cleavage.

The fragments of human DNA and the plasmid vector are then added together in the presence of an enzyme called DNA ligase. This enzyme joins the complementary sticky ends together thus inserting the pieces of human DNA into the plasmid vector.

Host bacterium

The plasmid vector is added to a suspension of the host bacterium and various mechanisms may be employed to transport the DNA into the cell. These may include transduction (Chapter 13) or electroporation. At the end of the process a small number of cells will have acquired the plasmid vector. Of those, only a small number will contain foreign DNA, and of those only a very small number will contain the gene of interest.

Selection

The cells can be plated onto agar containing ampicillin and only those cells that have acquired the plasmid vector (containing the resistance gene) will be able to grow.

Screening

The vector is often designed such that the restriction site is located within the β-galactosidase gene. Sometimes the vector may be modified so that cells containing plasmid with an intact β-galactosidase gene produce blue colonies on appropriate media while those that have a defective β-galactosidase gene due to insertion of the foreign DNA will produce white colonies on the same medium. The white colonies can therefore be picked off and grown up individually so that they can be further screened for the production of the gene of interest.

20.4 Microbial biotransformation of steroids

Steroids occur naturally in the body and possess a wide array of pharmacological properties. Examples include:

- adrenal corticosteroids (cortisone, corticosterone);
- progestational hormones (progesterone);
- androgens (testosterone);
- oestrogens (oestradiol, oestrone).

They are all derived from same basic ring structure (see Figure 20.1), and variations arise by changing the substituents on this ring. Over the last 60 years hundreds of synthetic variants have been produced giving rise to a variety of molecules with the potential to be:

- anti-inflammatory drugs;
- sedatives;
- antitumour drugs;
- cardiovascular drugs;
- oral contraceptives;
- dermatological agents.

20.4.1 Historical context

In 1949 it was found that cortisone had remarkable anti-inflammatory properties when injected for the treatment of rheumatoid arthritis. This news led to a tremendous demand for the drug as there was no alternative satisfactory treatment for this condition at that time, but since the cortisone had to be extracted from natural sources the demand could not be met. There was no chemical method for the large scale manufacture of cortisone in those days.

A chemical method was designed later which used deoxycholic acid as a starting material and this could be extracted from cattle bile. However, 31 separate chemical steps were required to produce cortisone; the yield was poor and the process economically unsound. For

x

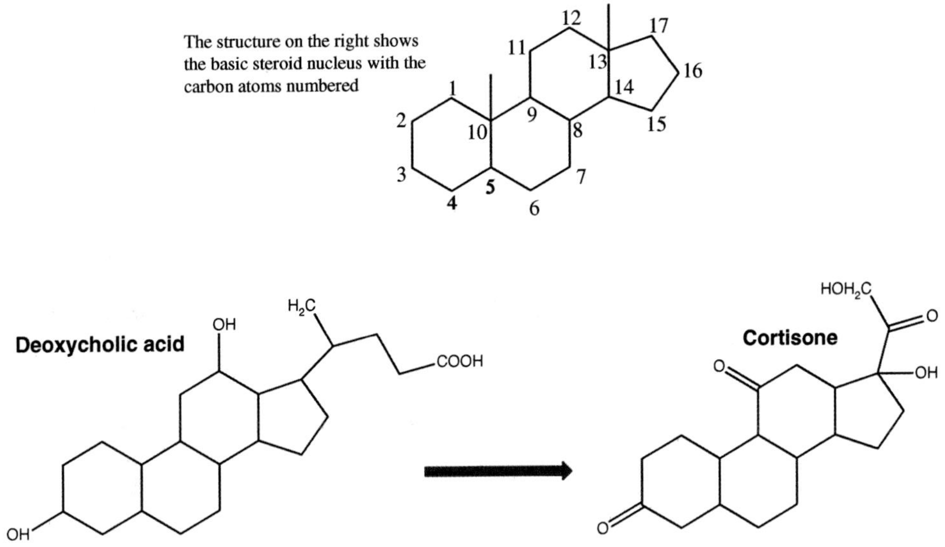

The structure on the right shows the basic steroid nucleus with the carbon atoms numbered

Deoxycholic acid

Cortisone

Figure 20.1 The basic steroid nucleus and the first example of the chemical synthesis of cortisone.

example, over 600 kg of deoxycholic acid were needed to produce 1 kg of cortisone. Figure 20.1 shows the oxygen shift from C12 to C11, which is essential for activity, but this move alone required nine separate chemical steps.

A more efficient production process was therefore required, which used cheaper and more plentiful supplies of starting materials. It was also necessary to reduce reliance on chemical synthetic pathways for the more complex reactions. In the first instance it was found that yams were a source of diosgenin and soybeans were a source of stigmasterol (both natural plant products containing the steroid nucleus). Independently it was observed that bacteria and fungi could carry out oxidation, isomerization and hydroxylation reactions on the steroid nucleus and hence these two discoveries led to an alternative approach to the production of therapeutic steroids.

20.5 Microbial modifications of the steroid nucleus

Rhizopus arrhizus (*nigricans*) was found to carry out 11α hydroxylation of the steroid nucleus at C11 and this had a greater than 85% yield. In addition, *Curvularia lunata* carried out 11β hydroxylation; *Streptomyces argenteolus* carried out the 16α hydroxylation and *Streptomyces lavendulae* introduced a 1,2 double bond. Figure 20.2 shows that from the more plentiful starting materials

described above a combination of microbial and chemical processes can give rise to a range of steroidal agents with high yield and purity.

20.6 Primary metabolic products

Primary metabolites are typically small molecules, which arise as a result of the normal growth and metabolism of the cell. Figure 20.3 illustrates the production of primary metabolites during the stage of active growth of the culture. Rapid growth and high cell concentrations give rise to high yields of primary metabolite. In most cases the cells in the wild produce very little of the product we are interested in and the amounts are regulated by various feedback mechanisms.

An example is shown in Figure 20.4 where the desired product B is obtained by the action of enzyme 1 on compound A. The production of enzyme 1 is controlled via a negative feedback loop where, as the concentration of B increases, the level of enzyme 1 decreases. In the example shown here the picture is further complicated because our desired product is further metabolized in the cell to compound C via the action of enzyme 2. The levels of enzyme 1 may also be controlled by the concentration of compound C. If we wish to maximize the production of compound B then we must select strains of cultures in which these control mechanisms have been removed or greatly reduced. It would be necessary to block the negative feedback via compounds B and C and also to

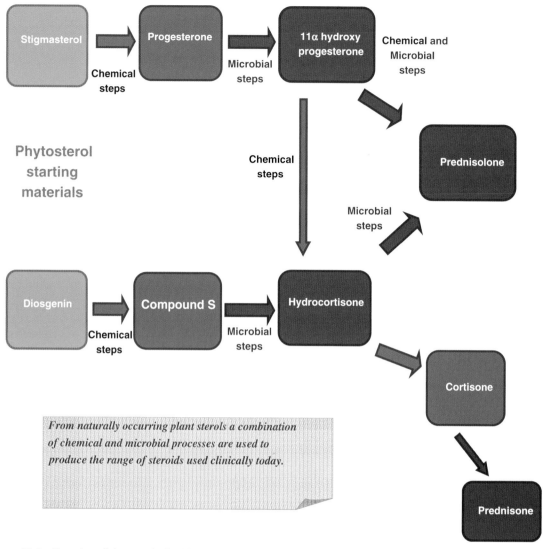

Figure 20.2   Examples of the use of microbial biotransformations in steroid synthesis.

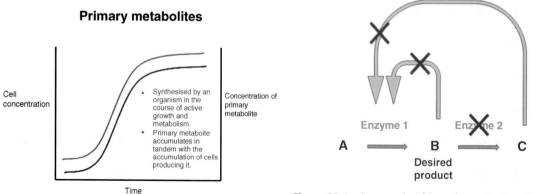

Figure 20.3   Primary metabolites (blue line) are produced during active growth of microorganisms (shown by the red line).

Figure 20.4   An example of how the production of a useful compound must be manipulated to maximize yields (details given in the text).

Table 20.3 Pharmaceutical uses of amino acids produced by microorganisms.

L-Amino acid	Uses
Arginine	Infusions, cosmetics
Glutamine	Assists recovery from trauma; TPN
Histidine	TPN
Isoleucine	TPN
Leucine	Dietary supplement; TPN
Lysine	TPN
Methionine	TPN
Phenylalanine	Nutritional supplement (supposed analgesic & antidepressant); TPN
Proline	Osmoprotectant in pharmaceutical formulations
Serine	Cosmetics
Tryptophan	Dietary supplement (antidepressant activity); TPN
Tyrosine	Mood modifier? L-DOPA synthesis
Valine	TPN

TPN = total parenteral nutrition.

inhibit the production of enzyme 2. Such cultures are derived through a process of mutation and selection and in this way organisms which hugely overproduce compounds of commercial interest are obtained. Examples of these highly mutated organisms include:

- *Ashbya gossypii* produces 20 000 times more riboflavin (B$_2$) than the natural wild type.
- *Propionibacterium shermanii* and *Ps. denitrificans* produce 50 000 times more cobalamin (B$_{12}$) than their naturally occurring counterparts.

We have already seen that primary metabolites include alcohols, amino acids, organic acids, nucleotides and vitamins. Table 20.3 gives examples of various amino acids produced by microorganisms and their uses in the pharmaceutical industry.

20.7 Secondary metabolic products

Secondary metabolites are also typically small molecules and, most importantly, include antibiotics. Figure 20.5 illustrates the production of secondary metabolites in

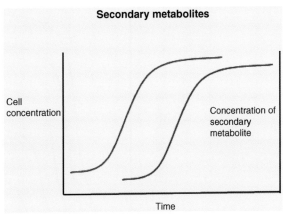

Secondary metabolites

Cell concentration

Concentration of secondary metabolite

Time

Figure 20.5 Secondary metabolites (blue line) do not accumulate in the medium until after the cells have stopped growing (shown by the red line).

relation to growth of the culture. As can be seen, the compound of interest is not produced until the cells enter stationary phase, and unlike primary metabolites high growth rates inhibit yield. In order to maximize yields we need to have slow growth but high cell concentrations. Achieving this combination is not straightforward because high cell concentrations require nutritious media which in turn give high rates of growth.

Let us take a simple medium which has glucose as its sole source of carbohydrate and this is the ingredient which runs out first during growth. If this is inoculated with a small number of cells then those cells will grow at their maximum rate until the glucose is exhausted and the culture will enter stationary phase. If we now reduce the glucose concentration somewhat and repeat the process the rate of growth will be the same but the final cell concentration will be less because there is less carbon to manufacture cells. We can repeat this by gradually reducing the concentration of glucose and each time the growth rate remains at a maximum but the final cell concentration gets gradually less (see Figure 20.6). Eventually, the glucose concentration is so low that the cells cannot grow at their maximum rate and the culture grows more slowly, however, the final cell yield is now very poor.

If we require low glucose concentrations to achieve slow growth but high glucose concentrations to achieve high cell numbers then the answer may be to formulate the medium with a high glucose concentration but to feed the glucose into the culture very slowly throughout growth. This is known as an open system of culture compared to a closed (batch) culture where all the ingredients are present at the start. Figure 20.7 illustrates this process which is used to maximize yields.

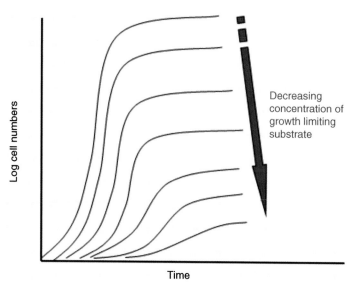

Figure 20.6   Decreasing the concentration of growth limiting substrate initially does not affect growth rate – only the final cell yield is reduced. As the concentration is reduced further both rate of growth and yield are reduced.

20.8 Commercial production of antibiotics

Antibiotics are secondary metabolites but it is important to realize that they are not waste products.

They are elaborate molecules constructed in the cell via a number of complex steps. In nature they are produced in very low amounts but industrially the cells are mutated such that they vastly overproduce the antibiotics. The production of antibiotics is now a multibillion dollar industry.

20.8.1 Penicillin

Penicillin was discovered by Alexander Fleming in 1928 at St. Mary's hospital, Paddington. He was studying the bacterium *Staphylococcus aureus* which required regular examination of his agar plates over several days but his agar plate became contaminated with a mould culture from the air. The important observation made by Fleming was *not* that there was a zone of inhibition, as that is quite a common occurrence, but that the colonies of bacteria had become established and were then subsequently lysed by the mould. Fleming tried without

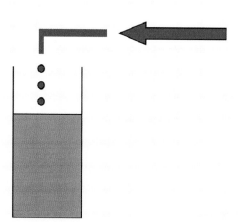

Cells are inoculated into an incomplete rich medium (contains all components necessary for growth except one – e.g. glucose).

Glucose is then added slowly to the culture and as the concentration in broth is low, growth is slow.

By the end of the process the same amount of glucose will have been added as would have been present in the complete rich medium.

Figure 20.7   Glucose is added slowly to a growing culture to force the microorganisms to grow slowly.

success to isolate the active ingredient from mould cultures and this was eventually achieved by Florey and Chain in 1939. Norman Heatley developed a suitable production process in 1941.

Fleming described penicillin production using *Penicillium notatum* and this was used for the initial development work. However, yields were poor and the mould only produced antibiotic when growing on the surface of agar. Subsequently, *P. chrysogenum* was tested and found to give good yields in liquid culture.

Early clinical use of penicillin

The first clinical application of penicillin was in 1930 when Cecil Paine, a pathologist at the Royal Infirmary in Sheffield, attempted to use crude culture filtrates topically to treat patients with sycosis barbae (infection of the hair follicles on the face) but this was not successful due to lack of skin penetration. However, he did successfully treat an infant with ophthalmia neonatorum (gonococcal infection of the eyes) and this was the first recorded cure for penicillin.

The first human patient to receive purified intravenous penicillin was Albert Alexander, in 1941. He had developed bacteraemia from a cut hand whilst gardening and responded well to the novel therapy. However, stocks of the new drug were limited and there wasn't sufficient to complete the course. He died shortly afterwards.

Further improvements were made to the production process by altering the growth medium. Originally simple, standard media such as Czapek Dox agar were used and these were supplemented randomly to try to increase yields. Two changes that had profound effects were the use of lactose instead of glucose, which greatly increased yields due to slower growth, and the addition of corn-steep liquor, a byproduct from the wet milling of corn. The latter increased yields fivefold because it is a source of phenylacetic acid, a component of the penicillin molecule.

With *P. chrysogenum* as the starting point, a programme of strain mutation and selection then followed using mutagens such as nitrogen mustard, X-rays and UV light. The process of mutation and strain selection is

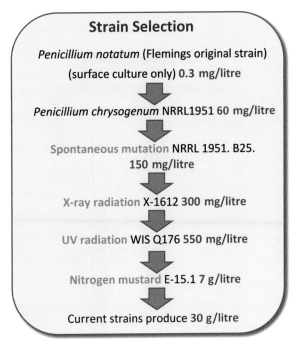

Figure 20.8 Improvement of penicillin-producing strains through mutation and selection.

summarized in Figure 20.8 and shows the final industrial strains of *P. chrysogenum* producing about 30 mg per ml of culture.

20.8.1.1 The fermentation process

The industrial production of penicillin takes place in large stainless steel fermentation vessels of about 180 000 litres capacity. A diagram of a typical fermenter is shown in Figure 20.9. These vessels are stirred rapidly and aerated with forced sterile air. They are fitted with temperature, pH and foam control and pumps for the slow administration of additional nutrients.

The process flow chart (Figure 20.10) shows that the fermenter is inoculated from a seed tank of 500 litres capacity and after the fermentation is complete the penicillin is contained within the cell-free medium. This means that the culture can be filtered to remove the cells and the supernatant treated to extract the antibiotic. To extract the penicillin the solution is acidified; this allows the antibiotic to partition into a solvent such as amyl or butyl acetate. This stage must be performed rapidly and at low temperature as the penicillin is unstable in acidic solutions. Addition of phosphate buffer causes the penicillin to crystallize out where it can be

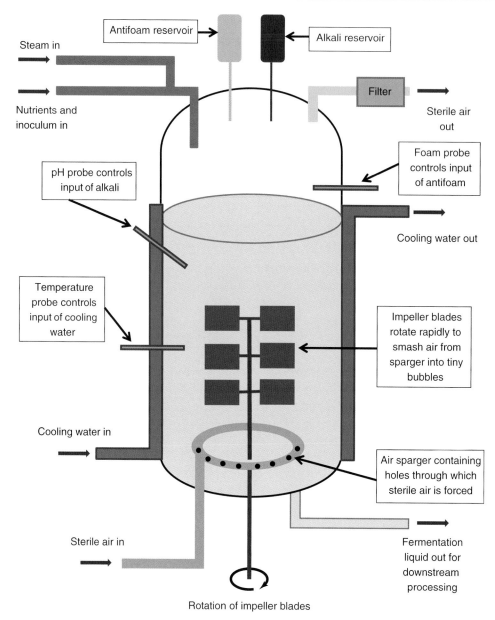

Figure 20.9   Schematic diagram of a large-scale fermentation vessel.

washed and dried. If the culture is able to produce 30 mg per ml of penicillin, a fermenter capacity of 180 000 litres will give a total yield of penicillin per batch of around 5400 kg. With a dose of 250 mg each batch therefore yields 20 million doses.

The final product of the industrial fermentation process is benzyl penicillin (penicillin G) but, although this is effective, it has a number of limitations. Its spectrum of activity is restricted to Gram-positive bacteria with little or no activity against Gram-negative bacteria. It is acid labile and is therefore destroyed by gastric acid in the stomach thus requiring it to be administered parenterally. Finally, it is readily inactivated by β-lactamases leading to the development of resistance (Chapter 10).

Chemical analysis of fermentation broths always gives higher concentrations than biological analysis. The difference is the presence of 6-amino penicillanic acid (6-APA) which is a precursor of benzyl penicillin but has no antimicrobial activity. 6-APA is benzyl penicillin without the side chain and it is the nature of the −R

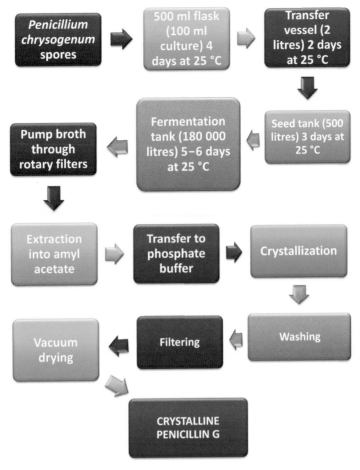

Figure 20.10   Penicillin fermentation process flow chart.

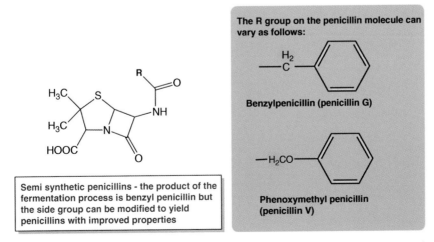

Figure 20.11   The basic penicillin nucleus illustrating how the side chain may be manipulated to produce products with different characteristics.

group, which determines the characteristics of the penicillin molecule. 6-APA is therefore a useful starting material for making penicillin molecules with different properties (see Figure 20.11).

It is not practical to manufacture 6-APA directly as it is much more difficult to extract from the fermentation broths; therefore benzyl penicillin is manufactured and converted to 6-APA using microbial enzymes. After extraction and purification it is possible to chemically add different side chains to give penicillin molecules with improved properties. Thus the production of

penicillin is known as a semisynthetic process where the cells carry out part of the process and the remainder is carried out chemically.

Acknowledgement

Chapter title image: Alexander Fleming – discoverer of penicillin. http://commons.wikimedia.org/wiki/File: Alexander_Fleming.jpg

Index

Essential Microbiology for Pharmacy and Pharmaceutical Science, First Edition. Geoffrey Hanlon and Norman Hodges.
© 2013 John Wiley & Sons, Ltd. Published 2013 by John Wiley & Sons, Ltd.